T0256830

Statistical Analysis of
Geographical Data

Statistical Analysis of Geographical Data

An Introduction

Simon J. Dadson

School of Geography and the Environment,
University of Oxford, UK

This edition first published 2017
© 2017 John Wiley and Sons Ltd

All rights reserved. No part of this publication may be reproduced, stored in a retrieval system, or transmitted, in any form or by any means, electronic, mechanical, photocopying, recording or otherwise, except as permitted by law. Advice on how to obtain permision to reuse material from this title is available at http://www.wiley.com/go/permissions.

The right of Simon J. Dadson to be identified as the author of this work has been asserted in accordance with law.

Registered Offices
John Wiley & Sons Ltd, The Atrium, Southern Gate, Chichester, West Sussex, PO19 8SQ, UK

Editorial Office
9600 Garsington Road, Oxford, OX4 2DQ, UK

For details of our global editorial offices, customer services, and more information about Wiley products visit us at www.wiley.com.

Wiley also publishes its books in a variety of electronic formats and by print-on-demand. Some content that appears in standard print versions of this book may not be available in other formats.

Limit of Liability/Disclaimer of Warranty

While the publisher and authors have used their best efforts in preparing this book, they make no representations or warranties with respect to the accuracy or completeness of the contents of this book and specifically disclaim any implied warranties of merchantability or fitness for a particular purpose. No warranty may be created or extended by sales representatives or written sales materials. The advice and strategies contained herein may not be suitable for your situation. You should consult with a professional where appropriate. Neither the publisher nor authors shall be liable for any loss of profit or any other commercial damages, including but not limited to special, incidental, consequential, or other damages.

Library of Congress Cataloging-in-Publication Data

Name: Dadson, Simon J.
Title: Statistical analysis of geographical data : an introduction / Simon J. Dadson.
Description: 1 edition. | Chichester, West Sussex : John Wiley & Sons, Inc., 2017. | Includes bibliographical references and index.
Identifiers: LCCN 2016043619 (print) | LCCN 2017004526 (ebook) | ISBN 9780470977033 (hardback) | ISBN 9780470977040 (paper) | ISBN 9781118525111 (pdf) | ISBN 9781118525142 (epub)
Subjects: LCSH: Geography–Statistical methods. | BISAC: SCIENCE / Earth Sciences / General.
Classification: LCC G70.3 .D35 2017 (print) | LCC G70.3 (ebook) | DDC 910.72/7–dc23
LC record available at https://lccn.loc.gov/2016043619

Cover image: © Joshi Daniel / EyeEm/GettyImages
Cover design: Wiley

Set in 10/12pt Warnock by SPi Global, Pondicherry, India

10   9   8   7   6   5   4   3   2   1

Contents

Preface *xi*

1 **Dealing with data** *1*
1.1 The role of statistics in geography *1*
1.1.1 Why do geographers need to use statistics? *1*
1.2 About this book *3*
1.3 Data and measurement error *3*
1.3.1 Types of geographical data: nominal, ordinal, interval, and ratio *3*
1.3.2 Spatial data types *5*
1.3.3 Measurement error, accuracy and precision *6*
1.3.4 Reporting data and uncertainties *7*
1.3.5 Significant figures *9*
1.3.6 Scientific notation (standard form) *10*
1.3.7 Calculations in scientific notation *11*
Exercises *12*

2 **Collecting and summarizing data** *13*
2.1 Sampling methods *13*
2.1.1 Research design *13*
2.1.2 Random sampling *15*
2.1.3 Systematic sampling *16*
2.1.4 Stratified sampling *17*
2.2 Graphical summaries *17*
2.2.1 Frequency distributions and histograms *17*
2.2.2 Time series plots *21*
2.2.3 Scatter plots *22*
2.3 Summarizing data numerically *24*

2.3.1   Measures of central tendency: mean, median and mode   *24*

2.3.2   Mean   *24*

2.3.3   Median   *25*

2.3.4   Mode   *25*

2.3.5   Measures of dispersion   *28*

2.3.6   Variance   *29*

2.3.7   Standard deviation   *30*

2.3.8   Coefficient of variation   *30*

2.3.9   Skewness and kurtosis   *33*

        Exercises   *33*

3   Probability and sampling distributions   *37*

3.1   Probability   *37*

3.1.1   Probability, statistics and random variables   *37*

3.1.2   The properties of the normal distribution   *38*

3.2   Probability and the normal distribution: z-scores   *39*

3.3   Sampling distributions and the central limit theorem   *43*

        Exercises   *47*

4   Estimating parameters with confidence intervals   *49*

4.1   Confidence intervals on the mean of a normal distribution: the basics   *49*

4.2   Confidence intervals in practice: the *t*-distribution   *50*

4.3   Sample size   *53*

4.4   Confidence intervals for a proportion   *53*

        Exercises   *54*

5   Comparing datasets   *55*

5.1   Hypothesis testing with one sample: general principles   *55*

5.1.1   Comparing means: one-sample z-test   *56*

5.1.2   *p*-values   *60*

5.1.3   General procedure for hypothesis testing   *61*

5.2   Comparing means from small samples: one-sample *t*-test   *61*

5.3   Comparing proportions for one sample   *63*

5.4   Comparing two samples   *64*
5.4.1   Independent samples   *64*
5.4.2   Comparing means: *t*-test with unknown population
          variances assumed equal   *64*
5.4.3   Comparing means: *t*-test with unknown population
          variances assumed unequal   *68*
5.4.4   *t*-test for use with paired samples (paired *t*-test)   *71*
5.4.5   Comparing variances: *F*-test   *74*
5.5   Non-parametric hypothesis testing   *75*
5.5.1   Parametric and non-parametric tests   *75*
5.5.2   Mann–whitney *U*-test   *75*
          Exercises   *79*

**6   Comparing distributions: the Chi-squared test   *81*
6.1   Chi-squared test with one sample   *81*
6.2   Chi-squared test for two samples   *84*
          Exercises   *87*

**7   Analysis of variance   *89*
7.1   One-way analysis of variance   *90*
7.2   Assumptions and diagnostics   *99*
7.3   Multiple comparison tests after analysis
          of variance   *101*
7.4   Non-parametric methods in the analysis
          of variance   *105*
7.5   Summary and further applications   *106*
          Exercises   *107*

**8   Correlation   *109*
8.1   Correlation analysis   *109*
8.2   Pearson's product-moment correlation
          coefficient   *110*
8.3   Significance tests of correlation coefficient   *112*
8.4   Spearman's rank correlation coefficient   *114*
8.5   Correlation and causality   *116*
          Exercises   *117*

**9   Linear regression   *121*
9.1   Least-squares linear regression   *121*
9.2   Scatter plots   *122*

9.3   Choosing the line of best fit: the 'least-squares'
       procedure   *124*
9.4   Analysis of residuals   *128*
9.5   Assumptions and caveats with regression   *130*
9.6   Is the regression significant?   *131*
9.7   Coefficient of determination   *135*
9.8   Confidence intervals and hypothesis tests concerning
       regression parameters   *137*
9.8.1 Standard error of the regression parameters   *137*
9.8.2 Tests on the regression parameters   *138*
9.8.3 Confidence intervals on the regression
       parameters   *139*
9.8.4 Confidence interval about the regression line   *140*
9.9   Reduced major axis regression   *140*
9.10   Summary   *142*
       Exercises   *142*

10   **Spatial statistics**   *145*
10.1   Spatial data   *145*
10.1.1 Types of spatial data   *145*
10.1.2 Spatial data structures   *146*
10.1.3 Map projections   *149*
10.2   Summarizing spatial data   *157*
10.2.1 Mean centre   *157*
10.2.2 Weighted mean centre   *157*
10.2.3 Density estimation   *158*
10.3   Identifying clusters   *159*
10.3.1 Quadrat test   *159*
10.3.2 Nearest neighbour statistics   *162*
10.4   Interpolation and plotting contour maps   *162*
10.5   Spatial relationships   *163*
10.5.1 Spatial autocorrelation   *163*
10.5.2 Join counts   *164*
       Exercises   *171*

11   **Time series analysis**   *173*
11.1   Time series in geographical research   *173*
11.2   Analysing time series   *174*
11.2.1 Describing time series: definitions   *174*
11.2.2 Plotting time series   *175*

11.2.3 Decomposing time series: trends, seasonality and irregular fluctuations *179*
11.2.4 Analysing trends *180*
11.2.5 Removing trends ('detrending' data) *186*
11.2.6 Quantifying seasonal variation *187*
11.2.7 Autocorrelation *189*
11.3 Summary *190*
Exercises *190*

Appendix A: Introduction to the R package *193*
Appendix B: Statistical tables *205*
References *241*
Index *243*

Preface

Quantitative reasoning is an essential part of the natural and social sciences and it is therefore vital that any aspiring geographer be equipped to perform quantitative analysis using statistics, either in their own work or to understand and critique that of others. This book is aimed specifically at first year undergraduates who need to develop a basic grounding in the quantitative techniques that will provide the foundation for their future geographical research. The reader is assumed to have nothing more than rusty GCSE mathematics. The clear practical importance of quantitative methods is emphasized through relevant geographical examples. As such, the book progresses through the basics of statistical analysis using clear and logical descriptions with ample use of intuitive diagrams and examples. Only when the student is fully comfortable with the basic concepts are more advanced techniques covered. In each section, the following format is employed: (i) an introductory presentation of the topic; (ii) a worked example; and (iii) a set of topical, geographically relevant exercises that the student may follow to probe their understanding and to help build confidence that they can tackle a wide range of problems. Use of the popular R statistical software is integrated within the text so that the reader can follow the calculations by hand whilst also learning how to perform them using industry-standard open source software. Files containing the data required to solve the worked examples are available at https://simondadson.org/statistical-analysis-of-geographical-data.

I am grateful for the guidance and wisdom of my own academic advisers: Barbara Kennedy, who sadly died before the book was completed, Mike Church, and Niels Hovius. I am grateful to seven anonymous readers of the book's outline for their positive support for the idea. To that I must add further thanks owed to colleagues at Oxford University, the Centre for Ecology and Hydrology, and elsewhere for their help and encouragement during the writing of this book. Particular thanks go to Richard Field, Richard Bailey, Toby Marthews and Andrew Dansie who read earlier drafts of the manuscript and made many useful suggestions that have undoubtedly improved the style of the book. My special thanks go to the large number of undergraduate and graduate students who have read the chapters and worked through the exercises in this book. Of course, any remaining errors or ambiguities are my own and I would be most grateful to have them brought to my attention.

At Wiley, I owe a considerable debt of gratitude to Fiona Murphy for her encouragement to undertake this project, Rachael Ballard for commissioning the work, and to Lucy Sayer, Fiona Seymour, Audrie Tan, Ashmita Thomas Rajaprathapan, Wendy Harvey and Gunal Lakshmipathy for their diligence in helping to see the work through to completion.

Finally, I would like to thank my wife, Emma, and our two children, Sophie and Thomas, for their support throughout the process of writing this book and for their tolerance of the time it has taken. To them this book is dedicated.

Oxford, 2016   *Simon J. Dadson*

1

Dealing with data

STUDY OBJECTIVES

- Understand the nature and purpose of statistical analysis in geography.
- View statistical analysis as a means of thinking critically with quantitative information.
- Distinguish between the different types of geographical data and their uses and limitations.
- Understand the nature of measurement error and the need to account for error when making quantitative statements.
- Distinguish between accuracy and precision and to understand how to report the precision of geographical measurements.
- Appreciate the methodological limitations of statistical data analysis.

1.1 The role of statistics in geography

1.1.1 Why do geographers need to use statistics?

Statistical analysis involves the collection, analysis and presentation of numerical information. It involves establishing the degree to which numerical summaries about observations can be justified, and provides the basis for forming judgements from empirical data.

Statistical Analysis of Geographical Data: An Introduction, First Edition.
Simon J. Dadson.
© 2017 John Wiley & Sons Ltd. Published 2017 by John Wiley & Sons Ltd.

Take the following media headlines, for example:

We know in the next 20 years the world population will increase to something like 8.3 billion people.
  Sir John Beddington, UK Government Chief Scientist[1]
2010 hits global temperature high.
  BBC News, 20th January 2011[2]

Each of these statements invites critical scrutiny. The reliability of their sources encourages us to take them seriously, but how do we know that they are correct? It is hard enough to try to predict what one human being will do in any particular year, let alone what several billion are going to do in the next 20 years. How were these predictions made? How was the rate of change of world population calculated? What were the assumptions? What does the author mean by 'something like'? The number 8.3 billion is quite a precise number: why didn't the author just say 8 billion or almost 10 billion?

Similarly, how do we know that 2010 is the *global* temperature high, when temperature is only measured at a small number of measuring stations? How would we go on to investigate whether anthropogenic warming caused the record-breaking temperature in 2010 or whether it was just a fluke?

Statistical analysis provides some of the tools that can answer some of these questions. This book introduces a set of techniques that allow you to make sure that the statistical statements that you make in your own work are based on a sound interpretation of the data that you collect.

There are four main reasons to use statistical techniques:

- to describe and measure the things that you observe;
- to characterize measurement error in your observations;
- to test hypotheses and theories;
- to predict and explain the relationships between variables.

---

[1] http://www.bbc.co.uk/news/science-environment-12249909.
[2] http://www.bbc.co.uk/news/science-environment-12241692.

1.2   About this book

One of the best ways to learn any mathematical skill is through repeated practice, so the approach taken in this book uses many examples. The presentation of each topic begins with an introduction to the theoretical principles: this is then followed by a worked example. Additional exercises are given to allow the reader to develop their understanding of the topics involved.

The use of computer packages is now common in statistical analysis in geography: it removes many of the tedious aspects of statistical calculation leaving the analyst to focus on experimental design, data collection, and interpretation. Nevertheless, it is essential to understand how the properties of the underlying data affect the value of the resulting statistics or the outcome of the test under evaluation.

Two kinds of computer software are referred to in this book. The more basic calculations can be performed using a spreadsheet such as Microsoft Excel. The advantages of Excel are that its user interface is well-known and it is almost universally available in university departments and on student computers. For more advanced analysis, and in situations where the user wishes to process large quantities of data automatically, more specialized statistical software is better. This book also refers to the open-source statistical package called 'R' which is freely available from http://www.r-project.org/. In addition to offering a comprehensive collection of well-documented statistical routines, the R software provides a scripting facility for automation of complex data analysis tasks and can produce publication-quality graphics.

1.3   Data and measurement error

1.3.1   Types of geographical data: nominal, ordinal, interval, and ratio

Four main types of data are of interest to geographers: nominal, ordinal, interval, and ratio. *Nominal data* are recorded using categories. For example, if you were to interview a group of people and record their gender, the resulting data would be on a nominal, or categorical, scale. Similarly, if an ecologist were to categorize the plant species found in an area by counting the number of individual plants observed in different categories, the resulting dataset would be categorical, or nominal. The distinguishing property of nominal data is that the categories are simply names – they cannot be ranked relative to each other.

Observations recorded on an *ordinal scale* can be put into an order relative to one another. For example, a study in which countries are ranked by their popularity as tourist destinations would result in an ordinal dataset. A requirement here is that it is possible to identify whether one observation is larger or smaller than another, based on some measure defined by the analyst.

In contrast with nominal and ordinal scale data, *interval scale* data are measured on a continuous scale where the differences between different measurements are meaningful. A good example is air temperature, which can be measured to a degree of precision dictated by the quality of the thermometer being used, among other factors. Whilst it is possible to add and subtract interval scale data, they cannot be multiplied or divided. For example, it is correct to say that 30 degrees is 10 degrees hotter than 20 degrees, but it is not correct to say that 200 degrees is twice as hot as 100 degrees. This is because the Celsius temperature scale, like the Fahrenheit scale, has an arbitrarily defined origin.

Ratio scale data are similar to interval scale data but a true zero point is required, and multiplication and division are valid operations when dealing with ratio scale data. Mass is a good example: an adult with a mass of 70 kg is twice as heavy as a child with a mass of 35 kg. Temperature measured on the Kelvin scale, which has an absolute zero point, is also defined as a ratio scale measurement.

It is important from the outset of any investigation to be aware of the different types of geographical data that can be recorded, because some statistical techniques can only be applied to certain types of data. Whilst it is usually possible to convert interval data into ordinal or nominal data (e.g. rainfall values can be ranked or put into categories), it is not possible to make the conversion the other way around.

1.3.2   Spatial data types

Geographers collect data about many different subjects. Some geographical datasets have distinctly *spatial* components to them. In other words, they contain information about the location of a particular entity, or information about how a particular quantity varies across a region of interest. In many contexts, it is advantageous to collect information on the locations of objects in space, or to record details of the spatial relationships between entities. The two main types of spatial data that can be used are vector data and raster (or gridded) data. *Vector data* consist of information that is stored as a set of points that are connected to known locations in space (e.g. to represent towns, sampling locations, or places of interest). The points may be connected to form lines (e.g. to represent linear features such as roads, rivers and railways), and the lines may be connected to form polygons (e.g. to represent areas of different land cover, geological units, or administrative units).

The locations of points must be given with reference to a coordinate system which may be rectangular (i.e. given using eastings and northings in linear units such as metres), or spherical (i.e. given using latitudes and longitudes in angular units such as degrees), but which always requires the definition of unit vectors and a fixed point of origin. The most common spherical coordinate system is that of latitude and longitude, which measures points by their angular distance from an origin which is located at the equator (zero latitude) and the Greenwich meridian (zero longitude). Thus the latitude of Buckingham Palace in London, UK, is 0.14°W, 51.50°N indicating that it is 0.14 degrees west of Greenwich and 51.5 degrees north of the equator.

Whilst spherical coordinate systems are commonly used in aviation and marine navigation, and with the arrival of GPS, terrestrial navigation usually uses rectangular coordinate systems. In order to use rectangular coordinates, the spherical form of the Earth must be represented on a flat surface. This is achieved using a map projection. An example of a map projection that is used to obtain a rectangular coordinate system is the Great Britain National Grid, in which locations are defined in metres east and north of a fixed origin that is located to the south west of the Scilly Isles. Thus to give a grid reference for Buckingham

Palace as (529125, 179725) is to say that it lies at a point which is 529.150 km east of the origin and 179.750 km north of the origin. To reduce the amount of information that must be transmitted in practical situations, grid references are typically given relative to a set of predefined 100 km squares. In situations where quoting distances to the nearest metre is not justified they are usually rounded to a more suitable level of precision. The grid reference above, for Buckingham Palace, might be rounded to the nearest 100 m and associated with the box TQ [which has its origin at (500000, 100000)] to give TQ 291 797, where two letters indicate the grid square, the first three digits indicate the easting and the last three digits indicate the northing.

Raster data are provided on a grid, where each grid square contains a number that represents the value of the data within that grid square. Almost any kind of data can be represented using a raster. Examples of data that are collected in raster format include many types of satellite image, and other datasets that are sampled at regular intervals (see Section 2.1.3). The technical process of specifying the location of the raster in space is identical to the process used to locate a point, described above. It is also necessary to specify the resolution of the raster (i.e. the spacing between grid points and the extent or size of the domain).

1.3.3 Measurement error, accuracy and precision

All measurements are subject to uncertainties. As an example, consider a geographer wishing to measure the velocity of a river. One way to do this is to use a stopwatch to measure the time it takes a float to travel a known distance. What are the uncertainties involved in this procedure? One source of error is the reaction time of the person using the stopwatch: they might be slow starting the watch, or fast stopping the watch, or vice versa. Since each possibility is equally likely, this kind of error is termed *random error*. One way to measure the amount of random error in a measurement is to repeat the procedure many times: sometimes the time will be underestimated, other times we will overestimate the time. By analysing the variability or spread in our results, we can get a good estimate of the amount of random error in our observation. If the spread is small, we say that our measurement is precise; if the spread is large, our measurement is less precise. The term *precision* is used to describe the degree to which repeated observations of the same quantity are in agreement with each other.

What if the stopwatch was consistently slow? In this case, all of the times measured would be shorter than they ought to be and no amount of repetition would be able to detect this source of error. Such errors are referred to as *systematic* errors, because we consistently underestimate the time taken if the stopwatch is slow, and consistently overestimate the time taken if the stopwatch is fast. If the amount of systematic error is low, we refer to our measurements as accurate; if the amount of systematic error is high, our measurement is less accurate. The term *accuracy* is used to describe the degree to which a measured value of a quantity matches its true value. Statistical analysis offers few opportunities to detect systematic errors, because we do not usually know the true value of the measurement that is being made: it is up to the person measuring the data to reduce the amount of systematic error through careful design of field, lab, or survey procedures.

A typical graphical analogy used to illustrate the difference between accuracy and precision involves a set of archery targets (Figure 1.1). Here, the archer is subject to random errors due to the wind or the steadiness of their hand; and potential systematic errors due to the design of the bow and arrow and its sight. Note the important point that it is impossible to assess the precision of a single measurement using statistical techniques.

1.3.4 Reporting data and uncertainties

The most straightforward way to communicate error is to give the best estimate of the final answer and the range within which you are confident that the measurement falls. Taking the earlier example of measuring the velocity of a river, suppose that we measure the velocity several times, giving the following estimates (in metres per second, or m/s):

0.5, 0.4, 0.5, 0.6

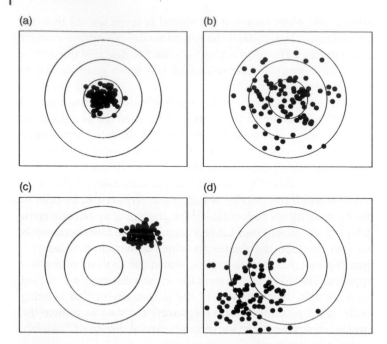

Figure 1.1 Accuracy and precision in archery. (a) High accuracy with high precision; (b) high accuracy with low precision; (c) low accuracy but high precision; (d) low accuracy and low precision. Note that without knowing the location of the archery target (i.e. the true value of the measured quantity), cases (c) and (d) are indistinguishable from (a) and (b), respectively.

The best estimate of the river's velocity is the average of these measurements, which is 0.5 m/s (calculated by adding up all of the values and dividing by the number of values). What about the range? At the most basic level, it makes sense to assume that the correct answer lies between the lowest (0.4 m/s) and the highest (0.6 m/s) values. We will want to refine this approach later on but for now we can say that the best estimate is 0.5 m/s ± 0.1 m/s, or to put it more generally:

$$\text{Measured value} = \text{best estimate} \pm \text{error}$$

(1.1)

Note that the '±' symbol is pronounced 'plus-or-minus'. In many situations, for data measured on a ratio scale, it is useful to express the error as a percentage of the measured value. So 0.5 ± 0.1 m/s would become $0.5 \pm 20\%$ m/s.

It is worth noting that the example above, in which we used the range of observed values to estimate the error, is the simplest approach available. One of the aims of the statistical analyses described in this book is to provide more advanced ways to quantify this error, including the use of confidence intervals based on probabilities.

1.3.5 Significant figures

It is unwise to claim that your measurements are more precise than they really are. For example, it would be misleading to state that the average age of a group of interviewees was 23.357 if each person's age were known only to the nearest whole number. Rounding to an appropriate precision can be achieved by looking at the number of significant figures in the data.

The first significant figure (sig. fig. or s.f.) is the first figure which is not zero (reading from the left). For example, the first significant figure in the following numbers is indicated with a box:

$\boxed{1}$25, 0.0$\boxed{1}$25, 0.0000$\boxed{1}$25

The number of significant figures can then be counted from left to right, ignoring embedded zeros. The following examples all have four significant figures (boxed):

$\boxed{1205}$, $\boxed{12.05}$, 0.$\boxed{1205}$, 0.00$\boxed{1205}$

To round a number to:

1 s.f. look at the 2nd s.f.
2 s.f. look at the 3rd s.f.
3 s.f. look at the 4th s.f.

If the digit that you look at is less than 5, ignore it and round down. If it is 5 or more, round up by adding one to the digit in front.

You should round the final answer to reflect the precision of the original data. In general, the last significant figure quoted in your final result should be of the same order of magnitude as the error. It is important to remember that you should only round the results of calculations as the last step in the chain of calculations. This is essential because otherwise you may incur round-off errors in the intermediate steps of the calculation.

1.3.6 Scientific notation (standard form)

Sometimes, it is necessary to use numbers that are very large or very small. For example, the amount of carbon stored in the terrestrial biosphere is enormous (approximately 600 000 000 000 000 000 g), whilst the size of a typical cell is tiny (about 0.00001 m). It is inconvenient to have to write down all these zeros. Scientific notation (or standard form) provides a concise way to report such numbers. The general definition of a number in scientific notation is: $a \times 10^n$, where a is a real number between 1 and 10, and n is an integer. So the numbers above convert to:

$$600\,000\,000\,000\,000\,000\,\text{grams} = 6.0 \times 10^{17}\,\text{g}\left(= 600\,\text{Pg}\right)$$

$$0.000\,01\,\text{m} = 1.0 \times 10^{-5}\,\text{m}\left(= 10\,\mu\text{m}\right)$$

The shorthand prefixes for various powers of 10 are given in Table 1.1.

Remember that the negative exponent refers to the reciprocal of the quantity concerned: $10^{-3} = 1/10^3$. This fact is sometimes used when writing down common units: so metres per second, or m/s, becomes m s^{-1}.

To convert a number into standard form, it is necessary first to write down the part of the number between 1 and 10 and then work out the power. For example, the number of people living in the UK is approximately 62 million people. To convert this to scientific notation, we have:

$$62\,\text{million} = 62\,000\,000$$

$$= 6.2 \times 10\,000\,000$$

$$= 6.2 \times 10^7$$

So, 62 million is 6.2×10^7. Note that computer programs and calculators often use a different way to display numbers in scientific notation, so in Microsoft Excel, for example, you might see 6.2×10^7 written as 6.2E + 07. Here the 'E' stands for '...times ten to the power of...'.

Table 1.1 Prefixes used to indicate powers of 10.

Number	Scientific notation	Prefix	Symbol
1 000 000 000 000 000 000	10^{18}	Exa-	E
1 000 000 000 000 000	10^{15}	Peta-	P
1 000 000 000 000	10^{12}	Tera-	T
1 000 000 000	10^9	Giga-	G
1 000 000	10^6	Mega-	M
1 000	10^3	Kilo-	k
1	1	—	—
$\dfrac{1}{1\,000}$	10^{-3}	Milli-	m
$\dfrac{1}{1\,000\,000}$	10^{-6}	Micro-	µ
$\dfrac{1}{1\,000\,000\,000}$	10^{-9}	Nano-	n
$\dfrac{1}{1\,000\,000\,000\,000}$	10^{-12}	Pico	p
$\dfrac{1}{1\,000\,000\,000\,000\,000}$	10^{-15}	Femto-	f
$\dfrac{1}{1\,000\,000\,000\,000\,000\,000}$	10^{-18}	Atto-	a

1.3.7 Calculations in scientific notation

Calculations using scientific notation are sometimes easier than with ordinary numbers, especially if the quantities are particularly large or small. The basic procedure is to group the numbers and the powers of 10, then work out the multiplications or divisions using the laws of indices (i.e. multiplication requires the addition of indices; division requires their subtraction).

Worked Example 1.1   The population of England and Wales in the 2011 census was estimated to be 56 075 900 people. The total land area of England and Wales is 245 000 km^2. Express each of these figures in standard form retaining all their precision and then find the population density in England and Wales to 3 significant figures.

Solution   First, to convert the figures into standard form (scientific notation) and round them to 3 s.f., the population becomes 56.0759 × 10^6 (or 56.0759 million) and the total land area is 2.45 × 10^5 km^2. To divide the population by the land area we write that:

$$\text{Population density} = \left(56.0759 \times 10^6\right) \div \left(2.45 \times 10^5\right)$$
$$= \left(56.0759 \div 2.45\right) \times \left(10^6 \div 10^5\right)$$
$$= 22.8881 \times 10^1$$
$$= 229 \text{ people per square kilometre, to 3 s.f.}$$

Exercises

1   The population of Northern Ireland in the 2011 census was 1 810 900 and the land area of Northern Ireland is 13 840 km^2. Express these numbers in scientific notation to 3 significant figures and calculate the population density in Northern Ireland.

2   Define the terms accuracy and precision and explain how you would quantify them in a geographical study of your choice.

3   The average annual rainfall in the catchment of the Thames to Kingston is 720 mm per year. The area of the Thames catchment draining to this point is 9948 km^2. The average discharge of the Thames at Kingston is 78 m^3/s. What fraction of the rain falling in the Thames catchment travels to the gauging station at Kingston?

2

Collecting and summarizing data

STUDY OBJECTIVES

- Appreciate the range of possible sampling procedures and the importance of randomization and replication in research design.
- Recognize a range of different graphical methods for presenting data (e.g. histograms, time series, scatter plots) and understand the circumstances in which each can be used.
- Understand the range of measures of central tendency: mean, median and mode.
- Appreciate some measures of dispersion: variance, standard deviation and inter-quartile range.

2.1 Sampling methods

2.1.1 Research design

In statistics the entire group of entities that is of interest is called the population, and it is desirable to be able to make statements about the population from a smaller fraction of the population, which is called a sample. Examples of geographical research in which sampling techniques are typically used include population census surveys, assessments of biodiversity from field samples, monitoring of atmospheric and oceanic processes using sparsely deployed instruments, and surveys and questionnaires designed to support interviews.

It is clear that in any study the results will be applicable only to the measurements made in that study, although it might be

Statistical Analysis of Geographical Data: An Introduction, First Edition.
Simon J. Dadson.
© 2017 John Wiley & Sons Ltd. Published 2017 by John Wiley & Sons Ltd.

hoped that they will lead to statements that can inform the development of a wider body of theory and that these inferences may then be subjected to more comprehensive tests. In any case it is important to define the target population at the outset of the study. The *target population* is the population that the study aims to investigate. In many cases it will be possible to enumerate the population, perhaps through an electoral roll, a list of shops in a shopping centre, or a set of trees in a forest. In other cases, the population will be harder to enumerate and in some situations it will be almost impossible to list the entire population (e.g. the number of fish in a river, the number of pebbles on a beach).

Once the target population is known, the *sampling frame* is the name given to the list of possible samples. The definition of the sampling frame is the first stage in sampling and even before any sampling of the population takes place a degree of judgement is necessary to determine whether the sampling frame is suitable. Providing that the sampling frame is chosen without bias, it is possible to make reliable inferences about the population based on the sample, which is one of the main reasons to use quantitative statistical analysis. However, the statistical techniques presented in this book can only quantify the random variability in the composition of the sample; they cannot correct for systematic biases in the selection of the sample in the first place.

One of most striking historical examples of hidden bias in the selection of a sampling frame occurred in the famous 1936 Presidential Election poll undertaken by *Literary Digest* magazine which erroneously predicted a decisive victory for Republican candidate Alfred Landon with 55% of the vote against 41% to Franklin D. Roosevelt and 4% to Lemke (Squire, 1988). In the election, Roosevelt won with 61% of the vote.

The flawed survey design began with the selection of the sampling frame: questionnaires were sent to voters selected from telephone books and automobile club membership lists. Moreover, whilst an astonishingly high number of questionnaires were sent out (10 million), less than one-quarter of them were returned. The survey was flawed in two important ways: first, the 2.4 million voters who bothered to return the questionnaire may not have been a representative fraction of the total population; second, the use of telephone lists and automobile club membership lists would in 1932 have limited the survey to wealthier voters who may have be less likely to vote for Roosevelt. Taken together, these limitations meant that even though the sample was large it was heavily biased (Squire, 1988).

2.1.2 Random sampling

Randomization is a fundamental requirement imposed on statistical analyses. With a random sample, each member of the sampling frame must have a quantifiable probability of being selected, and whether that individual is selected must be due solely to chance. Here it is important to note the difference between a random sample and a purposive sample, in which individuals are chosen because the investigator feels that their inclusion would be of interest to the study. Such methods have their place, and are frequently used in qualitative research, but they do not predispose themselves towards study using the techniques presented in this book. Moreover, there is no quantitative way to use the results from studies without randomization to make inferences about the wider population, and it is not possible to quantify the error associated with conclusions drawn from such studies.

Some advantages of random sampling include speed, efficiency, and reduced costs, along with the guard against possible investigator bias (latent or otherwise). Random sampling also offers the possibility of calculating an estimate of the likely error due to the sampling procedure.

Practical methods for random sampling are straightforward in theory, although they may be time-consuming in practice. Each member of the sampling frame is enumerated and given a reference number. Random numbers can then be used to select members of the sampling frame to use in the sample. A typical example might be a list of students in a school, a list of residents on an electoral roll, or a list of firms in a town. Random numbers can be generated easily either by consulting a table of previously generated random numbers, or by generating them specifically for the study in question using a calculator or a computer. It is worth noting that, strictly speaking, random sampling should be performed with replacement so that if the same individual is randomly chosen twice they should be included in the sample twice. In practice this poses a number of difficulties and so it is usual for sampling to occur without replacement, that is, if an individual has already been chosen for inclusion in the sample, they will not be chosen again.

With random sampling, all parts of the sampling frame have an equal chance of selection. This simplicity can also lead to disadvantages: it may quite easily occur that all of the individuals selected happen to be men, or that all of the plants selected happen to be on south facing slopes. A possible problem that can arise quite frequently in geographical studies is that individuals in the sampling frame are clustered in space rather than distributed uniformly. Random sampling that is spatially uniform can therefore miss important features of the underlying population. Neither of these problems constitutes a complete failure of random sampling to provide useful information, but uneven coverage of the sampling frame may produce a sample that does not cover all aspects of the population that the analyst considers relevant.

2.1.3 Systematic sampling

Systematic sampling offers a convenient possible remedy to the problems of random sampling in that it guarantees uniform coverage of the sampling frame, through the selection of every fifth member of the sampling frame, for example. Defined more formally, a sampling interval, i, is chosen such that i is defined as the ratio of the size of the sampling frame to the size of the required sample. For example, to select 100 individuals from a sampling frame of 500 the sampling interval is $500/100 = 5$, which means that having chosen an initial element from the sampling frame every fifth subsequent element of the sampling frame should be selected for inclusion in the sample. Nevertheless there are two major problems with this approach: (i) the sample is not strictly speaking a random sample, because once the initial element and sampling frequency have been chosen the other members of the sample are uniquely determined and therefore have not been chosen independently; and (ii) unless the sampling frame is listed in random order in the first place, the regular spacing of the sampling interval might coincide with other periodic structure in the list, resulting in bias. An example of the latter problem would arise, for example, if sampling grain size systematically along a beach when periodic bedforms were present on the surface.

2.1.4 Stratified sampling

Stratified sampling combines elements of systematic and random sampling to achieve a result that, in some circumstances, is better than would be achieved by either method alone. Stratification involves partitioning the sampling frame into blocks based on some relevant variable and then sampling randomly within these blocks. The blocks may be defined based on attributes of the data, or in geographical studies may be used to achieve spatial stratification. A good example of the former case for stratification would be to avoid any inherent bias towards male or female respondents by subdividing potential survey candidates into male and female groups and sampling a fixed number of individuals randomly within these two groups. Similarly, in order to achieve spatial stratification in a study of vegetation types, the study area could be divided into gridded units then sub-plots of vegetation could be sampled randomly within these units. The majority of the standard statistical techniques described in this book are applicable to random samples, and providing the variability within each stratum of a stratified design is similar it poses no problem to use the standard statistical formulae. The specific instances where stratified sampling is particularly appropriate are noted in Chapter 7, in the context of a procedure termed 'analysis of variance'.

2.2 Graphical summaries

A graphical representation of a dataset can give a clear indication of its main features. Different kinds of data call for different kinds of graphs, and the method of presentation will depend on the question being posed. Computer packages such as Excel and R make the task of producing a graph more straightforward, but they do not remove the need to consider how best to plot the data in order to address the question at hand.

2.2.1 Frequency distributions and histograms

Histograms show the frequency distribution of a dataset. Construction of a frequency distribution requires that each data point be placed into one of a number of different classes or bins.

The number (or frequency) of data points falling within each class is then plotted on a histogram using bars whose areas are proportional to the observed frequencies.

In order to calculate the frequency distribution, it is necessary to give some thought to the selection of class widths: too many and there will be lots of classes with no data; too few and the plot will lack meaning. Trial and error is the most effective way to select the number of classes; between 5 and 20 is usual, with a good rule-of-thumb being to use the square root of the number of observations as a guide to the number of classes. Often equal class widths are chosen, but if there are extreme outliers present in the dataset then it can be sensible to use unequal class widths. If unequal class widths are used, it is very important to ensure that the bars of the histogram are plotted with their widths proportional to the class widths so that the area of each bar remains proportional to the observed frequency of data in its class.

The *relative frequency* distribution can also be calculated; this is obtained by dividing the observed frequency in each class by the total number of observations. The *cumulative frequency* distribution provides another convenient description of the data, which can be thought of as a fraction of the data that lies below a given value and is useful in comparing different distributions.

As a diagnostic plot, the histogram illustrates some key features of the shape of a distribution (Figure 2.1). The symmetry or otherwise of a histogram indicates whether a dataset is skewed (i.e. asymmetric with a longer tail in one direction). The term 'positive skew' is used when the long tail extends towards values higher than the average (Figure 2.1a), whilst the term 'negative skew' is used when the long tail extends towards values lower than the average (Figure 2.1b). Histograms also reveal whether frequency distributions are flatter than usual (sometimes referred to as 'platykurtic'; Figure 2.1c), or more peaked than usual ('leptokurtic'; Figure 2.1d). The histogram can be used to extract from the dataset the modal class, which is the class that contains the most values. In many cases there will be a single modal class (as in Figure 2.1a–d) but in some cases there may be two distinct modal values (bimodality; e.g. Figure 2.1e), or the distribution may exhibit multimodality in which there are more than two distinct modal values (Figure 2.1f).

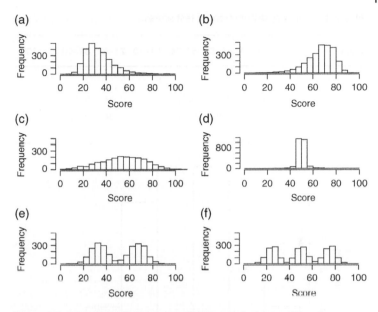

Figure 2.1 Examples of histograms showing typical properties of frequency distributions: (a) positive skew; (b) negative skew; (c) platykurtic; (d) leptokurtic; (e) bimodal; and (f) multimodal.

Table 2.1 Geography test scores.

96	89	82	78	78
74	73	72	67	66
64	64	63	61	60
60	59	59	58	55
50	49	49	39	34

Worked Example 2.1 The data shown in Table 2.1 give students' scores in Geography. Calculate the frequency distribution, relative frequencies and cumulative distribution for these data and plot a suitable histogram.

Solution Choosing appropriate classes so that class boundaries coincide with integer test score values gives the frequency distribution in Table 2.2. The relative frequency and cumulative

Table 2.2 Frequency distribution for test scores.

Class	31–40	41–50	51–60	61–70	71–80	81–90	91–100
Frequency	2	3	6	6	5	2	1
Relative frequency	0.08	0.12	0.24	0.24	0.20	0.08	0.04
Cumulative relative frequency	0.08	0.20	0.44	0.68	0.88	0.96	1.00

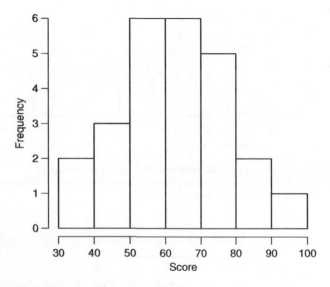

Figure 2.2 Histogram of test score data.

relative frequency have also been calculated in this table. The histogram itself (Figure 2.2) is straightforward to plot because the class widths are the same size.

Note that to use Excel for this task, first enter the data into the spreadsheet, in one column. To produce a histogram in Excel: Data > Data Analysis > Histogram (note that this requires the built-in Excel Analysis Toolpack to be activated using the 'Add-ins' menu). Set the input range to cover the column containing test score data. Leave the bin range for now, but specify the output range to be a point in your worksheet (this will be the top left-hand corner of where Excel will put its output). Tick the chart output box, to tell Excel to draw a chart.

Alternatively, you can use R to produce a similar plot. To read the data into R from a comma-separated variable (csv) file use the `read.csv()` function:

```
scores <- read.csv("wex2_1.csv")
```

To create a histogram, the `hist()` function is used:

```
hist(scores$Geography, breaks = 5,
     main-"Geography Test Scores",
     xlab="Score", ylab="Frequency")
```

Note that after specifying the data for which we would like the histogram (note from Appendix A that `scores$Geography` selects the column labelled 'Geography' from the data frame named 'scores'), a number of options (or arguments) control the way the histogram is plotted. The number of classes is controlled by setting `breaks` = 5 (although R will determine its own classes if you do not specify a value here). The title and axis labels can also be specified. The most frequently occurring values are close to the average value, which is about 60, and the distribution is fairly symmetrical.

2.2.2 Time series plots

Time series data, in which each value corresponds to an observation made at a particular point in time, occur very frequently in geographical applications. Such datasets are useful for exploring how different systems behave over timescales of various durations. It is often possible to investigate trends and seasonal cycles that would not be apparent otherwise. An example of a time series plot is given in Figure 2.3. This plot depicts a time series of monthly measurements of carbon dioxide (CO_2) made at the Mauna Loa Observatory and it is possible to distinguish not only the obvious trend over time, which represents an increase of approximately 50 parts per million (ppm) in CO_2 concentration over almost 40 years, but also the fluctuations in CO_2 concentration caused by seasonal differences in the respiration rates of terrestrial plants in the southern and northern hemispheres. Chapter 11 gives more information on specific techniques for describing and analysing time series data.

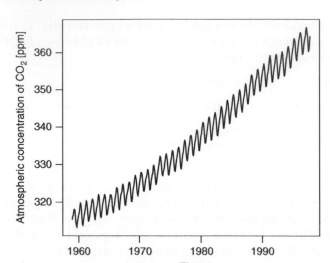

Figure 2.3 Time series of Mauna Loa atmospheric CO_2 concentrations. Note that this dataset is built into R and can be plotted using `plot(co2)`. Source: Keeling, C. D. and Whorf, T. P., Scripps Institution of Oceanography (SIO), University of California, La Jolla, California USA 92093-0220.

2.2.3 Scatter plots

Graphical methods for comparing datasets include the scatter plot in which pairs of values are plotted against each other. Such plots can easily be created in Excel and R. If two variables such as rainfall and crop yield are plotted, it can be seen that high values of July rainfall are associated with high corn yields (Figure 2.4). Techniques to quantify this association and to determine the likelihood of it being due to chance are discussed in detail in Chapter 8.

Where a dataset spans many orders of magnitude, it can be confusing to use a standard scatter plot. In such circumstances it is common to plot data on semi-log axes, where the increments in one axis are logarithmically spaced, which is to say that the axis labels go up in powers of ten rather than in a linear sequence as normal. The effect of using logarithmic axes can be seen in Figure 2.5a, which shows life expectancy at birth and income per person (gross domestic product or GDP per capita). Because many countries are bunched up in the low-income part of the graph on the left-hand side – that is, the data are skewed – the plot appears distorted. To get a better appreciation of the patterns in the dataset,

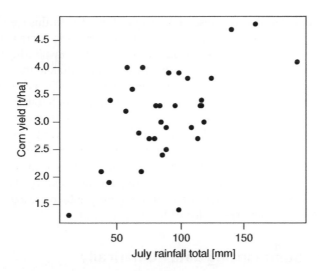

Figure 2.4 Scatter plot of July rainfall and corn yield in Iowa. Source: Data from United States Department of Agriculture – National Agricultural Statistics Service, http://www.nass.usda.gov.

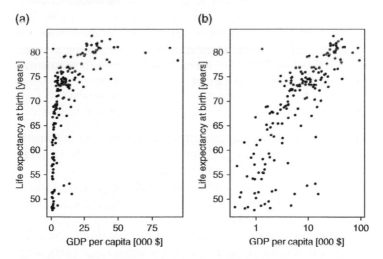

Figure 2.5 Life expectancy at birth and gross domestic product (GDP) per capita in dollars (adjusted for inflation): (a) with standard axes; and (b) with logarithmically spaced horizontal axis. Data: Gapminder – www.gapminder.org.

the same data can be presented on semi-log axes. In this case by constructing the horizontal axis with logarithmic increments (Figure 2.5b) the data are spread out more evenly. By plotting the graph in this way it is now possible to see more detail in the data. Of particular note is a subset of countries where GDP is around $10 000 per capita yet where life expectancy is much lower than would be expected. Clearly the analyst would look to investigate these countries further. It turns out that many of these nations have experienced civil wars in their recent history, or have economies dominated by natural resources, the revenues from which have not led to improved life expectancy for the majority of their inhabitants. In some circumstances it is appropriate to use log–log axes, where both axes are logarithmically spaced.

2.3   Summarizing data numerically

2.3.1   Measures of central tendency: mean, median and mode

Whilst graphical summaries give an intuitive overview of a dataset's properties, numerical summaries are essential when reporting results, making comparisons and conducting further statistical analyses. The most basic numerical summaries are simple measures of central tendency, which convey the 'average' value of the data. There are many different ways to summarize the central tendency of a dataset.

2.3.2   Mean

The most widely used is the arithmetic mean. The arithmetic mean is easy to calculate: it is simply the total of all the values in a dataset divided by the number of data values present.

Statisticians have a simple shorthand for this procedure. Suppose that we have a variable, x, which represents the measurement of interest (e.g. the height of a student). If there are three observations of this quantity we can put them in to a list and refer to them as x_1, x_2 and x_3. More generally, if there are n observations then they can be labelled x_i where i ranges from one to n. The shorthand for the sample mean is:

$$\bar{x} = \frac{\sum_{i=1}^{n} x_i}{n}.$$

(2.1)

Note that the upper-case Greek letter sigma, Σ, stands for 'sum' and indicates that you should add up all values of the quantity that follows, letting the index (in this case, i) vary from one to n. So to find the sample mean, $\bar{x}$ (read 'x bar'), you need to find the sum of all the values of x and divide by the number of observations, n.

It is important to note at this point that in most cases we do not have access to the full statistical population and so we are in fact using the sample mean to estimate the population mean (Section 2.1.1). The population mean is often given the symbol μ (the lower case Greek letter 'mu') and we are fortunate that, providing that the sample is random, unbiased and sufficiently large, the sample mean $\bar{x}$ is equal to the population mean, μ.

In addition to the arithmetic mean, two other kinds of mean are worth noting: the geometric mean is used when averaging percentage growth rates; and the harmonic mean is used when averaging rates of change.

2.3.3 Median

The median is another basic measure of central tendency. It represents the 'middle' value of a dataset and is most easily found by ranking the observations in order and choosing the one in the middle. If the number of values in the dataset is an odd number then selecting the middle value is easy; if there is an even number of values, then the arithmetic mean of the middle two values is taken. If the median is very close to the mean, it indicates that the data are not heavily skewed. The median is less sensitive to extreme values in the data than the mean and is often used as a more robust measure of central tendency when data do not follow a normal distribution.

2.3.4 Mode

The mode is the most commonly observed value in the data. It is the only measure of central tendency that can be used with data in categories or classes (i.e., nominal), when it can be helpful to state the most common category (or modal class), but it is not a very useful measure of central tendency when interval scale data are being used.

Worked Example 2.2   A geographer wishes to measure the effect of traffic pollution near to a major road. The geographer takes bark samples from pine trees in two sites: Site A is immediately next to the road, and Site B is 200 m away from the road. At each site a random sample of 24 trees is chosen. Bark samples from each tree are collected and the concentration of lead (Pb) is measured in ppm. The data are reported in Table 2.3. Find the mean and median lead concentration and use Excel or R to check your answer.

Solution   The mean is obtained by adding the data and dividing by the number of observations: For Site A, this gives 1574/24 = 66 ppm; for Site B the mean is 457/24 = 19 ppm. Note that in each case the answer must be rounded to 2 significant figures because the original data are measured to this precision.

The median value is obtained by ranking the data in numerical order, and finding the middle value. Since there is an even number of observations there is no single middle value so the arithmetic mean of the middle two values is taken. For Site A this gives 64 ppm and for Site B the median is 19 ppm. On either measure, the lead concentration at Site A appears to be substantially higher than that at Site B. The similarity of the mean and median for each site gives us an indication that neither dataset is highly skewed, although the histogram in Figure 2.6 is more conclusive.

To check this answer with Excel, enter the data and use the =average() function to calculate the mean and the =median() function to get the median. In R the data are read in using:

```
trees <- read.csv("wex2_2.csv")
```

and the mean is calculated using the mean() function:

```
mean(trees$A); mean(trees$B)
```

which gives the expected result (although we must round the answer by hand):

```
[1]  65.5
[1]  19.04167
```

Table 2.3 Concentration of lead (Pb) in pine trees (ppm).

Site A	Site B
69	23
51	26
69	17
59	20
79	14
55	12
70	27
64	20
63	16
81	15
58	23
75	18
56	18
67	26
62	24
55	12
48	21
75	20
89	15
77	16
68	22
62	17
60	20
60	15

The median is given by the median() function:

```
median(trees$A) ; median(trees$B)
```

which results in:

```
[1] 63.5
[1] 19
```

(a)

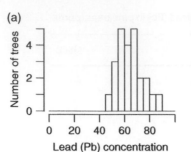

(b)

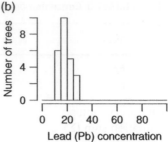

Figure 2.6 Histogram of lead concentration in pine trees at (a) Site A and (b) Site B.

Note that neither R nor Excel round the results to the correct number of significant figures automatically: that is a task that should still be completed by hand.

2.3.5 Measures of dispersion

In addition to summarizing the central tendency of a dataset it is also important to characterize its dispersion, or variability. Several measures can be used, including the range, which was defined as the difference between the maximum and minimum values in the dataset in Chapter 1. Because the range is sensitive to extreme values in the dataset (outliers) the inter-quartile range is often preferable, particularly when used with the median to provide summary of the data prior to further analysis. The inter-quartile range is calculated by ranking the dataset and then splitting it into four groups, each containing an equal number of observations. The divisions between these groups are termed quartiles (note that the second quartile is identical to the median), and the difference between the upper and lower quartiles gives the inter-quartile range.

The downside of using the inter-quartile range is that it does not incorporate the full range of information contained within the dataset. An alternative approach is to calculate the average difference between each data point and the mean value. Note that these differences will always add up to zero though, because the mean is by definition the central point in the data. Taking the absolute values of the differences (i.e. ignoring their plus or minus signs and finding the arithmetic mean of the absolute values of the differences) gives the average deviation, which is calculated as:

$$\text{average deviation} = \frac{\sum |x_i - \bar{x}|}{n} \qquad (2.2)$$

where $|x_i - \bar{x}|$ is the absolute value of the difference between each observation and the population mean. Note that when the summation sign (Σ) is given without limits explicitly stated it is assumed that the sum should be calculated using the entire dataset (i.e. with i = one to n).

2.3.6 Variance

An alternative approach, which is frequently used in statistics, is to find the mean of the *squared* differences. This gives the variance, which is given the symbol σ^2:

$$\sigma^2 = \frac{\sum (x_i - \mu)^2}{n} \qquad (2.3)$$

This formula is valid if data are available from an entire population, however, as was noted in Section 2.1.1, this is rarely the case; usually only a sample is available. If only a sample of the full population is available, then the sample variance must be calculated instead. Unlike the sample mean, which we noted in Section 2.3.2 was an unbiased estimator of the population variance, the estimation of the population variance from a sample needs a minor modification. The sample variance is calculated as:

$$s^2 = \frac{\sum (x_i - \bar{x})^2}{n-1} \qquad (2.4)$$

where it should be noted that the sample mean, $\bar{x}$, has been used instead of the unknown population mean, μ. This is an acceptable substitution, but only if we also use $n-1$ as the denominator in Equation 2.4. If we were to use the sample mean to calculate the sample variance without making this adjustment, we would on average underestimate the sample variance because for any particular sample the sample values will always tend to be closer to their particular sample mean than they will be to the population mean. To allow for this, we use $n-1$ as the denominator in our calculation of the sample variance.

Another way of putting this is to say that our use of the sample mean to calculate the variance has reduced the number of *degrees of freedom* that contribute to the calculation of the statistic. The n deviations must sum to zero, and so specifying any $n-1$ of them automatically determines the other one. In effect, for the calculation of the sample variance, we have only $n-1$ independent pieces of information: in other words, we have $n-1$ degrees of freedom. Thinking in terms of degrees of freedom gives us a useful shorthand for the formula for the variance, which is an important quantity that occurs in many areas of statistics:

$$\text{variance} = \frac{\text{sum of squares}}{\text{degrees of freedom}}.$$

2.3.7 Standard deviation

The standard deviation is obtained by taking the square root of the variance:

$$s = \sqrt{\frac{\sum(x_i - \bar{x})^2}{n-1}} \tag{2.5}$$

The standard deviation expresses the variability in the same units as the original data. It is an important measure of dispersion because in addition to describing the dispersion within a sample it has the useful property that whatever the underlying distribution of the data at least 75% of the observations lie within a range two standard deviations either side of the mean. This fact, which is a consequence of a rule known as Chebyshev's theorem, means that it is possible to think about probabilities associated with statements that we wish to make about the sample. In Chapter 3, we will make some more precise statements about the probabilities associated with statistical inferences.

2.3.8 Coefficient of variation

As noted above, the standard deviation has the same units as the original data from which it is calculated. This property brings the advantage that the standard deviation can be easily understood, but there are still some complications when comparing dispersion between datasets that have different means. As a solution to this problem the coefficient of variation can be used instead. This statistic is defined as the standard deviation divided by the mean:

$$CV = \frac{s}{\bar{x}}, \tag{2.6}$$

which is often expressed as a percentage. It should be noted that because the coefficient of variation involves division by the mean, it should be used only with ratio-scale data.

Worked Example 2.3 Calculate the variance, standard deviation and coefficient of variation of the lead concentrations in Table 2.3 and compare Sites A and B.

Solution The variance is calculated using Equation 2.4 by finding the difference between the mean and each data point, squaring it, adding this set of numbers up and dividing by the number of degrees of freedom. For Site A, the sum of squared differences between the sample values and the mean, $\sum (x_i - \bar{x})^2 = 2360$ as calculated in Table 2.4. Note that it is important to keep an extra significant figure in the calculations here, and the mean of 65.5 must be used instead of the rounded mean reported in Section 2.3.2, to keep the calculations precise. Dividing the corrected sum of squares by the number of degrees of freedom, which is $n-1 = 23$ because we have calculated one parameter (the mean), gives the variance, $s^2 = 102.6$. The standard deviation is the square root of the variance and so we have $s = 10.1$.

For Site B, the variance, $s^2 = 18.9$ and the standard deviation, $s = 4.34$. Thus the variability in Site A is more than twice that at Site B, although if we calculate the coefficient of variation for each site we get for Site A 0.15 and for Site B 0.23 which indicates that whilst the absolute amout of variation is higher at Site A the relative amount of variability is higher at Site B.

The function to calculate the variance in Excel is =var(); to calculate the sample standard deviation in Excel use =stdev(). There is an alternative function =stdevp() which calculates the standard deviation assuming that you have access to data from

Table 2.4 Calculation of variance and standard deviation for lead concentrations in pine trees.

Site A	$x_i - \bar{x}$	$(x_i - \bar{x})^2$
69	3.50	12.25
51	−14.5	210.25
69	3.5	12.25
59	−6.5	42.25
79	13.5	182.25
55	−10.5	110.25
70	4.5	20.25
64	−1.5	2.25
63	−2.5	6.25
81	15.5	240.25
58	−7.5	56.25
75	9.5	90.25
56	−9.5	90.25
67	1.5	2.25
62	−3.5	12.25
55	−10.5	110.25
48	−17.5	306.25
75	9.5	90.25
89	23.5	552.25
77	11.5	132.25
68	2.5	6.25
62	−3.5	12.25
60	−5.5	30.25
60	−5.5	30.25

n	24	Sum of squares	2360.0
$\bar{x}$	65.5	df	23
		Variance	102.6
		Std dev.	10.1

the entire population (i.e. with n in the denominator). The appropriate functions in R are var () and sd () which calculate the variance and standard deviation respectively.

2.3.9 Skewness and kurtosis

In Figure 2.1, histograms were shown which were not symmetric about the mean. These were described as 'skewed'. It is possible to quantify the degree of skewness in a sample numerically:

$$\text{skewness} = \sum \frac{\left(x_i - \bar{x}\right)^3}{\left(n-1\right)s^3} \tag{2.7}$$

The kurtosis or peakedness of the distribution can be quantified as:

$$\text{kurtosis} = \sum \frac{\left(x_i - \bar{x}\right)^4}{\left(n-1\right)s^4} \tag{2.8}$$

Measures of skewness and kurtosis are most useful in conjunction with the appropriate histogram when evaluating whether a dataset is drawn from a normal distribution (which has skewness of zero and kurtosis of 3). Many geographical datasets are highly skewed: variables such as income, city population, rainfall amounts and earthquake sizes are typically skewed, sometimes across many orders of magnitude. In such circumstances, the mean and standard deviation can, by themselves, give a misleading summary of the variable in question, because they are heavily influenced by the largest values in the dataset.

Exercises

1 Using the data in Table 2.3 plot histograms of lead concentration in pine trees similar to those shown in Figure 2.6.

2 Table 2.5 contains time series of rainfall observations in Glasgow and London. Plot the data using an appropriate method and comment on whether you see any trends, seasonal variability, or relationships between the two datasets. Calculate the mean and standard deviation of the annual rainfall amounts in each case.

Table 2.5 Rainfall data in Glasgow and London.

Year	Glasgow (mm)	London (mm)
1981	1237	680
1982	1343	650
1983	1174	535
1984	1103	642
1985	1340	526
1986	1438	637
1987	1120	654
1988	1284	552
1989	1192	534
1990	1533	425
1991	1225	533
1992	1394	649
1993	1255	614
1994	1424	631
1995	1241	539
1996	959	421
1997	1056	445
1998	1426	685
1999	1454	662
2000	1313	820
2001	843	705
2002	1346	787
2003	871	476
2004	1300	619
2005	1151	450
2006	1440	636
2007	1215	681
2008	1493	657
2009	1237	683
2010	947	521

3 A charity is interested in food poverty in a large city. The charity interviews 21 residents selected at random in each of two suburbs of the city and asks each of the residents to estimate their weekly food expenditure (Table 2.6). Calculate the mean, standard deviation and coefficient of variation of the two datasets.

Table 2.6 Weekly food expenditure in two suburbs of a city.

Parksville (£)	Skytown (£)
60	18
66	18
54	18
38	21
49	26
56	15
59	16
46	11
75	22
47	19
72	22
49	23
66	21
52	17
75	20
41	14
22	20
67	20
86	20
35	20
97	24

3

Probability and sampling distributions

STUDY OBJECTIVES

- Relate probabilities to values in a distribution.
- Understand how the central limit theorem allows us to approximate the sampling distribution for a statistic with a normal distribution and make inferences about the probable range in which population parameters lie.
- Know how to calculate the standard error of the mean and understand its significance.

3.1 Probability

3.1.1 Probability, statistics and random variables

Probability is the branch of mathematics concerned with matters of chance. Probabilistic statements are about the degree to which different uncertain outcomes are likely. For example, what is the chance of rain on Friday? How likely is there to be a recession? Most of the data analysis undertaken by geographers falls into three categories: (i) establishing the range of values that a statistic might reasonably take (e.g. what is the mean annual rainfall in London and how likely is it to vary from the average?); (ii) deciding whether there is a difference between two sampled values of a quantity of interest (e.g. is the annual income of residents in one part of a city significantly different from that of residents in another part?); and (iii) relating two or more variables to each other (e.g. how does mean annual rainfall relate to or crop production or umbrella sales?).

Statistical Analysis of Geographical Data: An Introduction, First Edition.
Simon J. Dadson.
© 2017 John Wiley & Sons Ltd. Published 2017 by John Wiley & Sons Ltd.

In situations where it is impossible to measure every individual in our population precisely, it is necessary to be able to calculate how reliably inferences about the population can be made using statistics calculated from a sample.

3.1.2 The properties of the normal distribution

To answer these questions, it is necessary to link the measures of central tendency and variability that have been presented in this book so far with the concept of probability. The normal distribution is especially important in this regard. It is bell-shaped and symmetrical about the mean (Figure 3.1). It can be defined by two parameters, the mean and standard deviation, and is often a good theoretical approximation for the true distribution of many of the measurements that geographers make.

The figure depicts the *probability density function* (PDF) for a standard normal distribution which has mean of zero and a standard deviation of one. This function expresses the likelihood of the quantity taking on any particular value shown. It is called the probability density because the probability of a value falling within a known range is given by calculating the area under the PDF over that range. The total area under the PDF is always equal to one.

The normal distribution illustrated in Figure 3.1 occurs frequently in geography and other empirical disciplines because many of the quantities that we measure have more than one influence. In order to obtain an extreme value, all of these influences must act in the same direction, but in almost all cases some of

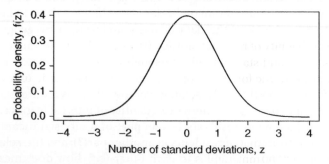

Figure 3.1 Theoretical form of the normal distribution.

them cancel each other out, giving a large number of values quite close to the mean and fewer towards the extreme tails of the distribution.

3.2 Probability and the normal distribution: z-scores

A normal distribution can have any mean and variance. Hence, to simplify matters it is useful to define a standard normal distribution, as a normal distribution that has a mean of zero and a standard deviation of one. Any normally-distributed quantity can be transformed so that it has a mean of zero and a standard deviation of one by subtracting the mean from each data value and dividing by the sample standard deviation. When a quantity is transformed in this way, it is called a z-score, or standardized value:

$$z_i = \frac{x_i - \mu}{\sigma}. \tag{3.1}$$

When data have been standardized, it is possible to see instantly how close an individual observation is to the mean.

One of the most important properties of the normal distribution is that 68% of its values lie within one standard deviation of the mean, and approximately 95% of its values lie within two standard deviations of the mean (Figure 3.2). It is possible to attach a particular probability to any given value covered by the normal distribution. Suppose that we are interested in the probability of certain rainfall values in a particular location. If it is known that the mean rainfall is 1000 mm per year with a standard deviation of 200 mm, we can pose questions such as: what is the probability of receiving over 1 m of rainfall in any given year? The mean and standard deviation can be used to calculate the z-score with the formula given in Equation 3.1, and the relevant probability can then be calculated.

To convert a z-score into a probability it is necessary to find out what fraction of the standard normal distribution lies to the right of that particular z-score. To take a simple example, the mean rainfall of 1000 mm has a z-score of zero, and we know that 50% of the distribution lies to the left of zero (Figure 3.3a). To find out

(a)

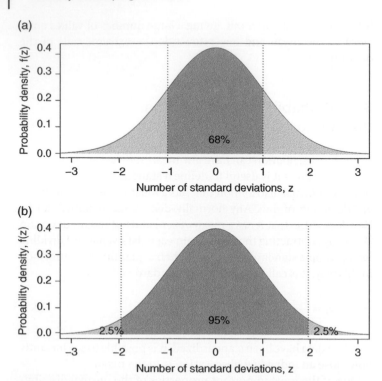

(b)

Figure 3.2 Properties of the normal distribution: (a) 68% of observations lie within one standard deviation either side of the mean; and (b) 95% of observations lie within two standard deviations either side of the mean.

what the probability is of receiving less than 800 mm of rainfall we can convert 800 mm into a z-score $(800 - 1000)/200 = -1.0$. We then need to find out what fraction of the possible values in the probability distribution lies to the left of a z-score of -1.0. From Figure 3.3b this is 16%. There is a 16% chance of receiving less than 800 mm of rainfall in any particular year. To take the example further, we could ask: what is the probability of receiving more than 1500 mm rainfall in any particular year. The z-score associated with 1500 mm is $(1500 - 1000)/200 = 2.5$. We know that 99.4% of the distribution of possible values lies to the left of a z-score of 2.5 (Figure 3.3c), so there is a 99.4% chance of receiving less than 1500 mm of rain. The question asked was what is the chance of receiving *more* than 1500 mm of rain though, which is

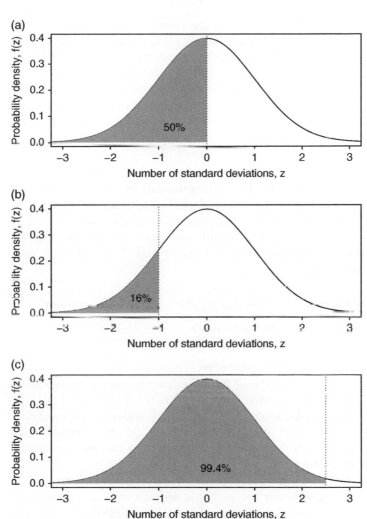

Figure 3.3 The area to the left of a particular z-score gives the fraction of the possible values in the distribution that are lower than that z-score. (a) z=0; (b) z=−1; (c) z=2.5.

0.6%, because the two probabilities must add up to one. This is indeed a very low probability (we might expect such a high annual rainfall amount to occur once in 167 years, and by that time the mean rainfall amount might have changed).

It is possible to compute the value for any z-score. To make things easier, many textbooks supply tables of probabilities associated with particular z-scores (see Appendix B, Table B.1), but it is also very straightforward to calculate the relevant probabilities directly using computer packages such as Excel, using the function = `normsdist()`, and in R using `pnorm()`. For example, =`normsdist(-1)` and `pnorm(-1)` each give the result 0.16, which tells us that 16% of the possible values in the distribution are lower than a z-score of −1.

Worked Example 3.1   Using the data in Table 2.5, calculate the probability of receiving less than 450 mm in any one year in London. Calculate the probability of receiving more than 1600 mm of rainfall in any one year in Glasgow. Explain the assumptions that you must make to arrive at your conclusions.

Solution   The mean and standard deviation are as follows:

London: $\bar{x}$ = 601.6; s = 101.3
Glasgow: $\bar{x}$ = 1245.1; s = 181.0

To find the probability of receiving less than 450 mm of rain in London in any one year, it is necessary to convert 450 mm into a z-score. For London, this is $(450 - 601.6)/101.3 = -1.50$. The probability associated with this z-score is given using = `norms-dist(-1.50)` in Excel or `pnorm(-1.50)` in R, both of which give 0.07 which means that there is only a 7% chance of such a low rainfall total in any particular year.

To find the probability of receiving more than 1600 mm of rain in Glasgow in any one year, it is necessary to convert 1600 mm into a z-score. For Glasgow, this is $(1600 - 1245)/181 = 1.96$. The probability associated with this z-score is given using = `norms-dist(1.96)` in Excel or `pnorm(1.96)` in R, both of which give 0.975 which means that there is a 97.5% chance of rainfall being less than 1600 mm or a 2.5% chance of rainfall being more than 1600 mm.

This analysis makes a number of assumptions, most importantly that the data follow a normal distribution, but also that the data are stationary (i.e. that there is no significant change over time).

3.3   Sampling distributions and the central limit theorem

The key step above was to use the z-scores to calculate probabilities so that we could say something about the chance of particular values occurring in the dataset. Yet the dataset we have is only one sample taken from the population as a whole. We could have taken a different sample, and if we had we might have got a different set of statistics. The value of the mean that was obtained from the sample would certainly be different had a different random sample been taken. This situation raises an important question: how can we know that our sample statistics are accurate and precise measures of the population statistics. The only way that we can control whether the statistics are accurate is to make sure that our sample is unbiased. If we interview only residents who live on the wealthiest streets we will get a biased estimate of the population of a town; if we measure sea surface temperature near the outlet of a power station, then we will get a biased estimate too. It is important to design the sampling strategy to overcome these biases. But even if we manage to design a good sampling strategy how can we deal with the fact that the same random sampling design could have led us to choose a different subset of the population? One way to investigate this effect would be to repeatedly take different random samples from the population and to calculate the distribution of their means. We would expect the distribution of means to be different, but we would hope that they would be centred around the true population mean. If the sampling design were unbiased then this would be the case, but Figure 3.4 illustrates that something more surprising also happens if we repeatedly calculate n means from a daily rainfall dataset. As we use more and more sample means to construct the histogram, the distribution of the averages tends closer and closer towards a normal distribution. This is an important and remarkable result; the daily rainfall values were themselves *not* normally distributed, but if our sample size is large enough, the distribution of the different means that we could have calculated *is* normally distributed. The theoretical distribution of the different means that we might have calculated had we chosen a different sample is called the

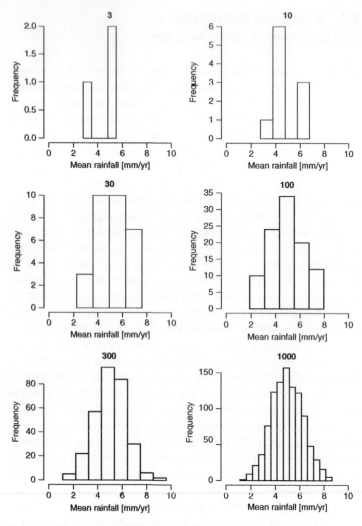

Figure 3.4 Illustration of the central limit theorem: the more samples we draw the closer their average approximates a normal distribution.

sampling distribution of the mean. We never usually see this distribution, but we can make some generalizations about it as a result of a mathematical theorem called the *central limit theorem,* which states that the mean of a large number of independent random variables will be normally distributed.

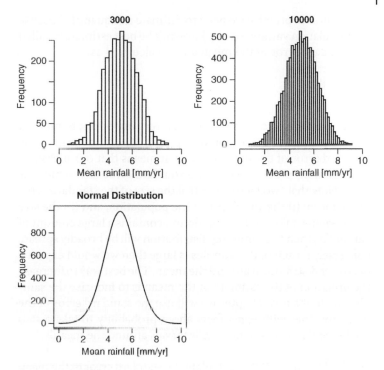

Figure 3.4 (*Continued*)

The mean of the sampling distribution will tend to be the same as the population mean (if it is not then our sampling method is biased), but what about the variability of the sampling distribution? The variance of the sampling distribution of the mean is:

$$\frac{\sigma^2}{n},$$

(3.2)

and the corresponding standard deviation is the square root of the variance:

$$\frac{\sigma}{\sqrt{n}}.$$

(3.3)

Of course in practice we need to estimate this quantity because the population variance is not known. The best estimate is called the standard error of the mean and is calculated as:

$$SE_{\bar{x}} = \frac{s}{\sqrt{n}} \qquad (3.4)$$

The standard error of the mean is a very important number. It is a measure of the precision of the estimate of the mean. If the standard error of the mean is small, it means that our measurement is more precise than if the standard error of the mean is large. Note that two factors control the size of the standard error of the mean: (i) the variability of the population; and (ii) the size of the sample taken. If the population contains a large amount of variability, then the sampling distribution will be broadly spread; conversely if each of the samples is large then we would expect a narrower distribution around the mean. The best way to increase the precision of the estimate of the mean is to increase the sample size. In the next chapter we will use the standard error of the mean together with some facts about probability to take a step further and look at confidence intervals around the mean.

Worked Example 3.2   Calculate the standard error of the mean rainfall in London and Glasgow and compare the two numbers.

Solution   To reiterate the calculations from Worked Example 3.1, we have:

London: $\bar{x}$ = 601.6; s = 101.3
Glasgow: $\bar{x}$ = 1245.1; s = 181.0
each with n = 30.

The standard error of the mean is therefore $SE_{\bar{x}} = \frac{101.3}{\sqrt{30}} = 18.5$ for London and $SE_{\bar{x}} = \frac{181.0}{\sqrt{30}} = 33.0$ for Glasgow. If we wanted to place a range of plausibility around our estimate of the mean, we might say that the mean was within two standard errors of the calculated value, giving for London $\bar{x} = 601 \pm 37$ and for Glasgow $\bar{x} = 1245 \pm 66$. It is interesting to note that these ranges do not overlap. This observation suggests that the two means are very different from each other, a theme which will be explored in more detail in Chapter 4.

Exercises

1 Calculate the probability of seeing a value lower than the following z-scores: $-3, -2, -1, 0, 1, 2, 3$ when drawing random samples from a standard normal distribution.

2 Another class of 25 students is ready to take the geography test described in Table 2.1. The pass mark for the test is 40. Assuming that the new students are drawn from the same population as the sample reported on in Table 2.1, what is the probability of a randomly selected student failing the test? The teacher proposes to set in advance a threshold for awarding a distinction. Any student scoring over the threshold will receive a prize. If the threshold is set at 80 marks, how many prizes should the teacher expect to give?

3 What is the standard error of the mean for the lead concentration measurements at the two sites reported in Table 2.3. Do you think that the difference between the two means is likely to have been a fluke?

4

Estimating parameters with confidence intervals

STUDY OBJECTIVES

- Construct a confidence interval around an estimate of the mean.
- Use a t-distribution to construct confidence intervals for small samples and when the population variance is unknown.
- Understand the issues involved in the choice of sample size.
- Construct confidence intervals around a proportion.

4.1 Confidence intervals on the mean of a normal distribution: the basics

In the previous chapter the standard error of the mean was defined as the standard deviation of the sampling distribution of the mean. We can use this fact to construct a confidence interval around the mean. With the help of the central limit theorem, we can say that had we taken many random samples and for each random sample we calculated a confidence interval which was two standard errors either side of the mean either side, approximately 96% of these confidence intervals would contain the true population mean, μ. More generally, the $100(1-\alpha)\%$ confidence interval can be calculated as:

$$\mu = \bar{x} \pm z_{\alpha/2} \frac{\sigma}{\sqrt{n}}, \tag{4.1}$$

Statistical Analysis of Geographical Data: An Introduction, First Edition.
Simon J. Dadson.
© 2017 John Wiley & Sons Ltd. Published 2017 by John Wiley & Sons Ltd.

where $z_{\alpha/2}$ is the upper $100\alpha/2$ percentage point of the standard normal distribution (Appendix B, Table B.1).

It is common in geographical and other studies to use a 95% confidence interval. We need to choose the range that encompasses 95% of the values in the distribution; this means we will leave out 5% of the distribution in the tails. There are two tails and so we need to leave out 2.5%, (i.e. $\alpha/2$) of the values in the distribution (shown in Figure 3.1). This value (2.5%) corresponds to 1.96 in Table B.1 and so we take a range which is 1.96 standard errors of the mean to construct the 95% confidence interval:

$$\bar{\mu} = x \pm 1.96 \frac{\sigma}{\sqrt{n}}, \qquad (4.2)$$

The precise meaning of the confidence interval is easiest to understand by recalling the thought experiment that was explained in Section 3.3. Here we imagined taking a large number of samples, and calculating the mean. If we extend this though experiment and imagine calculating the confidence interval as well, then we can expect that 95 times out of 100 the confidence interval will contain the true population mean. In practice, we usually only obtain one random sample and use it to calculate one confidence interval. It should be noted that this is not quite the same as stating that there is a 95% chance that the true population mean lies within the particular confidence interval that we have calculated, because in most circumstances we only obtain one random sample and only calculate one confidence interval.

4.2 Confidence intervals in practice: the *t*-distribution

The trouble with Equation 4.2 is that it requires us to know the population standard deviation in advance. In most cases the population standard deviation is not known. We could replace the population standard deviation with the sample standard deviation. For large sample sizes $(n > 30)$ this would be an appropriate approximate way to calculate the confidence interval, thanks to

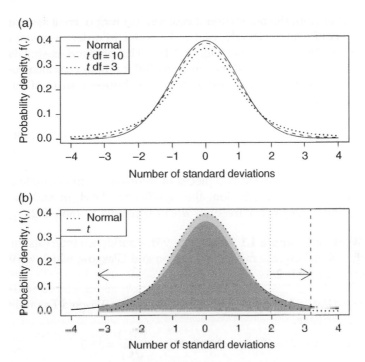

Figure 4.1 Comparison of *t*-distribution and normal distribution: (a) comparison of the normal distribution with the *t*-distribution for different numbers of degrees of freedom; and (b) area under the normal distribution compared with the area under the *t*-distribution.

the central limit theorem. However, the smaller the sample is the worse the approximation becomes. Fortunately, for small samples $(n < 30)$ and in situations where the population variance is not known beforehand, the *t*-distribution can be used instead (Figure 4.1).

The *t*-distribution is shown in Figure 4.1a, for various sample sizes. You can see that as the sample size gets larger the *t*-distribution becomes more and more like the normal distribution. When *n* is bigger than 30 the two distributions are virtually identical, but for small sample sizes the *t*-distribution is wider and flatter than the normal distribution (Figure 4.1a). What does this mean? It means two things: (i) when the sample size is small there is a higher risk of finding a sample mean that is further from the population mean; and (ii) with a small sample our confidence intervals need to be wider in order to contain 95% of the distribution (Figure 4.1b). The practical upshot is that when the sample size is small ($n < 30$) it is better to calculate the confidence intervals using the t-distribution instead of the z-distribution:

$$\mu = \bar{x} \pm t_{\frac{\alpha}{2}, n-1} \frac{s}{\sqrt{n}}, \qquad (4.3)$$

Note that you need two pieces of information to calculate a value of the t-distribution: the significance level, α; and the number of degrees of freedom, $n - 1$ (Appendix B, Table B.2).

Worked Example 4.1   Calculate 95% confidence intervals for the annual average rainfall in London and Glasgow, which were given in Table 2.5.

Solution   To reiterate the calculations from Worked Example 3.1, we have:

London: $\bar{x} = 601.6$; $s = 101.3$; $SE_{\bar{x}} = \dfrac{101.3}{\sqrt{30}} = 18.5$

Glasgow: $\bar{x} = 1245.1$; $s = 181.0$; $SE_{\bar{x}} = \dfrac{181.0}{\sqrt{30}} = 33.0$
each with $n = 30$.

Next, we need the value of the t-distribution with $\alpha = 0.025$ and $n - 1 = 29$ degrees of freedom. There are three ways to find this value: (i) with Excel [the function to use is = tinv(0.05,29), noting that Excel automatically divides the probability by two for you assuming you are interested in both tails of the distribution]; (ii) with R using qt(1 - 0.025,29), noting that the qt() function uses non-exceedance probabilities so we must subtract our α from 1; or (iii) using a table such as Table B.2. In each case the answer is that $t_{0.025, 29} = 2.045$ and so the size of the confidence interval either side of the mean is $2.045 \times 18.5 = 37.8$. We can say that the 95% confidence interval for the mean annual rainfall in London is 602 ± 38 mm (to the nearest mm).

For Glasgow, the standard error of the mean is 33.0. The value of the t-distribution with $\alpha = 0.025$ and $n - 1 = 29$ degrees of freedom is the same as it was before (2.045) and so the size of the confidence interval either side of the mean is 2.045 × 33.0 = 67.6. We can say that the 95% confidence interval for the mean annual rainfall in Glasgow is 1245 ± 68 mm (to the nearest mm).

Note that this does not mean that a value outside this range is abnormal. The interval given is the confidence interval on the mean, not the range of plausible values. It is also worth noting that the confidence intervals for London and Glasgow do not overlap. This indicates that the mean annual rainfalls at the two cities are significantly different, a very important conclusion that we will discuss in more detail in Chapter 5.

4.3 Sample size

The most effective way for a researcher to increase the statistical precision of a result is to increase the sample size. By rearranging the equation for the confidence interval, an approximate rule may be obtained that the sample size, n, that is required to ensure with $100(1 - \alpha)\%$ confidence that the difference between the sample and population means does not exceed a desired amount E is:

$$n = \left(\frac{z_{\alpha/2}\sigma}{E} \right)^2 \tag{4.4}$$

4.4 Confidence intervals for a proportion

In principle, a confidence interval can be found for any parameter estimate. A particularly common requirement is to compute the confidence interval on a proportion and a slightly different approach is required. If p is the proportion of observations in a random sample of size n that satisfy a certain criterion, then the standard error of the proportion (SE_p) is given as:

$$SE_p = \sqrt{\frac{p(1-p)}{n}} \tag{4.5}$$

which means that the proportion can be written alongside its confidence interval as:

$$p \pm z_{\alpha/2} \sqrt{\frac{p(1-p)}{n}} \tag{4.6}$$

where p is the true population proportion.

Worked Example 4.2 An exit poll of 150 students in an election shows that 98 voters chose Candidate A, and 52 voters chose Candidate B. What is the proportion of votes cast for Candidate A, and what is the confidence interval at 95%?

Solution The proportion of votes cast for Candidate A is $98/150 = 0.65$ (i.e. 65%). The standard error of this proportion is

$$\sqrt{\frac{p(1-p)}{n}} = \sqrt{\frac{(0.65)(0.35)}{150}} = 0.039, \text{ and } z_{\alpha/2} = 1.96,$$ the size of
the confidence interval is 0.076 or 8% (rounded to the nearest whole percentage point). We can therefore state that with 95% confidence the proportion lies within the range $65\% \pm 4\%$, or between 61% and 69%.

Exercises

1 Calculate 95% confidence intervals for the mean of the lead concentration measurements given in Table 2.3.

2 Find an appropriate confidence interval for the mean of the geography test scores in Table 2.1.

3 Calculate a confidence interval for the proportion of the electorate predicted to vote for F. D. Roosevelt in the 1936 US Presidential Election by the Literary Digest poll described in Section 2.1.1. What does your result tell you about the accuracy and precision of this poll?

5

Comparing datasets

STUDY OBJECTIVES

- Understand the basic principles of hypothesis testing and the key steps involved.
- Know when to use the Z-test and t-test for single sample hypothesis tests for means and proportions when comparing against a fixed value.
- Understand the meaning of p-values and appreciate the information that they contain.
- Perform two-sample tests with independent samples.
- Distinguish between two-sample t-tests for datasets with equal and unequal variances.
- Understand the difference between independent and paired sample tests and appreciate the benefits of paired sampling strategies where appropriate.
- Use simple tests for the equality of variances.
- Appreciate the difference between parametric and non-parametric tests for comparing datasets and understand which test should be used in which conditions.

5.1 Hypothesis testing with one sample: general principles

From simple measures of central tendency like the mean and the median, to the more complex calculation of sampling distributions and confidence intervals on the mean, the focus of the preceding chapters has been on quantifying error. More advanced

Statistical Analysis of Geographical Data: An Introduction, First Edition.
Simon J. Dadson.
© 2017 John Wiley & Sons Ltd. Published 2017 by John Wiley & Sons Ltd.

statistical techniques are required to make comparisons *between* different datasets, particularly in the presence of one or more hypotheses. Many different kinds of geographical questions can be posed as hypotheses. One might hypothesize that increasing carbon dioxide concentrations lead to higher global temperatures, or that educational attainment is higher amongst students taught in smaller classes. However, a statistical hypothesis is more specific: it is a claim about the parameters of one or more statistical populations. The set of steps that leads to a statement about whether a particular hypothesis is true is known as a hypothesis test.

5.1.1 Comparing means: one-sample z-test

The one-sample Z-test is used to test whether the mean of a single dataset conforms to a predetermined value. For example, consider a mineral water company which claims that its product contains 20 mg/l of magnesium. This quantity matters because magnesium is known to be important for its cardiovascular and other health benefits. We may test whether the company's claim is true by measuring the magnesium concentration in a sample of mineral water bottles. In statistical language, you are interested in testing the hypothesis that the magnesium concentration in the water is equal to 20 mg/l, and could write this as:

$$H_0 : \mu = 20 \, \text{mg} / l \qquad (5.1)$$

where μ is the population mean, as before. The symbol H_0 in Equation 5.1 stands for the null hypothesis and represents the situation where there is no difference from the specified value. The alternative hypothesis, H_1, covers the situation that there is a difference. The alternative hypothesis might be stated as:

$$H_1 : \mu \neq 20 \, \text{mg} / l \qquad (5.2)$$

We begin with the assumption that the null hypothesis is true and then look for convincing evidence that it is not. In the present case this means gathering data on magnesium concentrations. Suppose that we measure 50 samples of mineral water, chosen at random from different bottles, and find that the sample mean is 18.4 mg/l with a standard deviation of 4.9 mg/l. Should these data convince us that the null hypothesis is not true? On the face of it the sample mean is less than the specified value of 20 mg/l but someone wishing to scrutinize our study might point out that the standard deviation of 4.9 mg/l suggests a lot of variability between samples. It is important to note that both the null and alternative hypotheses refer to the population mean rather than the sample mean. We would not expect the sample mean to be precisely equal to the value stated in the null hypothesis, even if the null hypothesis were true. But how far from the stated value can our calculated sample mean lie before we do not feel confident that the population mean matches the predetermined magnesium concentration stated in the null hypothesis? In other words, if we were to assume for argument's sake that the mineral water company's stated claim that its water contains 20 mg/l, we can ask: 'what is the probability of finding the range of observed values that we obtained in our sample?' It is conventional to assume that this probability must be less than or equal to 0.05 (i.e. 5%) before we can declare that the null hypothesis is false. This probability is called the significance level, α. The corresponding confidence level is $1 - \alpha$ (or 0.95 or 95% in this example). It means that if the null hypothesis is true there is a 0.05 probability (or one-in-twenty chance) that we will reject it by mistake.

Rejecting the null hypothesis when it is true is called a Type I error. This type of error is sometimes also called a false positive, which means that the analyst has identified what they think is a significant effect when there is in fact none. Conversely, failing to reject the null hypothesis when it is false is called a Type II error. This type of error is often called a false negative, that is saying that there is no effect when in fact there is. The significance level, α, is a measure of the probability of making a Type I error, and one of the major goals of any statistical analysis is to minimize this probability. This is easy to do because analysts can select for themselves the value of α chosen when the test is performed. On the other hand, whilst we know that the probability of committing a Type II error (a false negative) decreases with the sample size, it is much harder to quantify. In practice, the particular situation tends to dictate the tolerable risk of making each type of error. A high level of α (the probability of a Type I error) might be chosen when the consequences of a false positive are more serious. For example, tests of an expensive new drug or a costly new environmental remediation strategy might demand more stringent restrictions on the probability of a Type I error or in other words a higher degree of statistical significance. In the case of a criminal conviction, the tolerability of making a Type II error (i.e. the risk of erroneously convicting an innocent person) is of huge importance, and must be considered accordingly.

How can the probability be calculated? One way is to calculate the difference between the measured sample mean and the magnesium concentration stated in the null hypothesis. In this case, the difference is 1.6 mg/l. This does not sound like a big difference, but remember that this sample mean is the result of having amalgamated information from fifty different individual measurements, and the larger our sample the more confidence we can have in our result. It makes no sense to look at the difference of 1.6 mg/l on its own; we need to see if this is a large difference compared with the amount of variability known to exist in the dataset. Dividing 1.6 mg/l by the standard error of the mean tells us how many standard errors this difference represents; it is also called the z-score:

$$Z_{calc} = \frac{\bar{X} - \mu_0}{\sigma / \sqrt{n}}, \tag{5.3}$$

where the numerator is the difference between the sample mean and the reference value, μ_0, against which the comparison is required. The denominator is the standard error of the mean. The z-score takes a normal distribution. Usually the population standard deviation is not known, but it is valid to assume that it is equal to the sample standard deviation unless the sample size is less than thirty (in which case a t-test is better, see Section 5.2). The value of Z_{calc} should then be compared with the critical value for a normal distribution (see Appendix B, Table B.1) to determine whether it falls within the 95% limits. If it does then the null hypothesis is accepted and the difference is judged not to be statistically significant; if Z_{calc} lies outside the acceptance region (i.e. in the critical region) then the null hypothesis is rejected and the difference is statistically significant.

The size of the critical region depends on the significance probability selected by the analyst at the outset but it is also partly controlled by the number of tails of the distribution in which the analyst is interested. If it is known in advance that the difference between the sample mean and the reference value, μ_0, against which the comparison is required is positive, then a one-tailed test will be sufficient. Similarly, if the analyst is interested only in knowing whether the difference is negative then a one-tailed test will suffice. However, if the question of whether the difference is positive or negative remains open (which is usually indicated by the alternative hypothesis) then the test should be two-tailed and the significance probability must be divided by two so that the probability is split evenly between the two tails of the distribution. In practice, one-tailed tests are quite rare – they can be used only if the analyst is already aware of the one-tailed nature of the hypothesis before the test is carried out, perhaps in the case of a regulatory pollution limit where it might be important to know if a pollutant value is significantly greater than the limit but of no interest to know if the value is significantly under the limit. If in doubt it is usually best to construct the null and alternative hypotheses to use a two-tailed test.

Worked Example 5.1 Perform a one-sample Z-test to determine whether the waters tested in Section 5.1.1 contained more or less than the stated magnesium concentration of 20 mg/l. Use a 5% significance level.

Solution
The calculated z-score is $Z_{\text{calc}} = \dfrac{18.4 - 20}{4.9 / \sqrt{50}} = \dfrac{-1.6}{0.69} = -2.32$, which means that the observed value is 2.32 standard errors lower than the null hypothesis value (i.e. the measured magnesium concentration was 2.32 standard errors lower than the claimed value). For a two-tailed test, if the null hypothesis is true then there is a $1 - \alpha$ probability that Z_{calc} lies within the critical region between $-z_{\alpha/2}$ and $z_{\alpha/2}$. For $\alpha = 0.05$, with two tails, the critical value is defined by $z_{\alpha/2} = z_{\alpha-0.025} = 1.96$ (see Table B.1), so we reject the null hypothesis if $Z_{\text{calc}} < -z_{\alpha/2}$ or if $Z_{\text{calc}} > z_{\alpha/2}$. Note that α is divided by two because this is a two-tailed test. In this case, $Z_{\text{calc}} = -2.32$, which is outside the critical region and so we can

reject the null hypothesis at the 0.05 significance level and conclude that there is a significant difference between the magnesium concentration in the mineral waters and the bottling company's stated claim.

5.1.2 *p*-values

The hypothesis-testing framework is unequivocal: once the analyst has selected a significance level it cannot (and should not) be changed afterwards. The downside of this approach is that it does not provide any information on how close the test came to rejecting the null hypothesis. In other words, it does not tell us how strong the evidence is against the null hypothesis. For this reason, it is now common to calculate and report the *p*-value associated with the test. This approach has several advantages: it indicates the strength of the evidence in favour of the null hypothesis; and it provides the reader with the means to set their own significance level if they are not happy with the one that the statistician has chosen.

For any calculated value of a test statistic, the *p*-value can most easily be thought of as the probability of obtaining that value of the test statistic or a more extreme value given that the null hypothesis is true. Alternatively, it can be thought of as the smallest α-level that would warrant rejection of the null hypothesis given the set of data under investigation.

Worked Example 5.2 Find the *p*-value associated with the test conducted in Worked Example 5.1.

Solution In the case of the Z-test, the *p*-value can be approximated using Table B.1 by finding the probability associated with the calculated value of the test statistic. In some cases, this is straightforward; in others it requires a level of approximation. In the present example, the *p*-value associated with a z-score of -2.32 is calculated by finding the area of the normal distribution which lies to the left of Z_{calc} (or to the right of Z_{calc} if it is positive). This can also be found easily using Excel with the function =normsdist(-2.27), or with R using pnorm(-2.32), both of which give the probability 0.01, which is the area to the left of the z-score. This probability should be doubled in the case of a two-tailed test in order to account for the upper tail too. Since the present test is a two-tailed test, the correct p-value is 0.02. In other words, there is only a 2% probability that the mineral water samples came from a population with magnesium concentration equal to 20 mg/l.

5.1.3 General procedure for hypothesis testing

In summary, the general procedure for hypothesis testing is to:

1) Identify the population parameter of interest.
2) State the null hypothesis and the alternative hypothesis.
3) Select a level of significance, α.
4) Decide whether the test is two-tailed or one-tailed.
5) Calculate the critical value(s) of the test statistic.
6) Calculate the test statistic.
7) Determine whether the null hypothesis should be rejected.
8) Report the results including the significance level (or the p-value) and the sample size.

By way of a practical note, it is worth stating that caution is often required in the interpretation of results from a hypothesis test when the number of samples is large because even a small difference between the hypothesized population mean can be counted as significant even when the difference is so small to be of little practical relevance.

5.2 Comparing means from small samples: one-sample *t*-test

When sample sizes are small ($n < 30$) or if the variance of the population is not known, a different procedure is required. This procedure is called a t-test and is sometimes referred to as Student's t-test, because its use was first documented by W. S. Gosset publishing under the name of Student. This test uses the t-distribution that was introduced in Section 4.2. For large samples ($n > 30$) the t-test gives similar results to the Z-test. It is an appropriate test to use to investigate differences between population means when the variance is unknown providing that the data are approximately normally distributed.

The principle of the test is identical to the Z-test described above, and the process of stating null and alternative hypotheses remains the same, but the test statistic, t_{calc}, is calculated as:

$$t_{calc} = \frac{\bar{X} - \mu_0}{s / \sqrt{n}}, \tag{5.4}$$

which is very similar to the definition of the z-score, except that the standard error has been calculated using the sample standard deviation, s. This test statistic has a t-distribution with $n-1$ degrees of freedom. The steps necessary to perform a t-test are almost identical to those required in a Z-test, but instead of using the normal distribution, the t_{n-1} distribution is used to define the critical region. As explained in Section 4.2, the t-distribution is flatter for smaller values of n, which means that a wider region is required to encompass the same proportion of the area under the curve. An important and desirable consequence of this property is that the smaller the sample size, the bigger the difference required in order to reject the null hypothesis.

Worked Example 5.3 Perform the comparison between the measured magnesium concentrations and their quoted value in Section 5.1.1 by using a one-sample t-test instead of a Z-test. Which is the better test to use in these circumstances and what are the consequences of this choice?

Solution To re-use the example above, the t-statistic is identical (i.e. $t_{calc} = -2.32$), the number of degrees of freedom is $50 - 1 = 49$, and so the critical value of the t-distribution at the $\alpha/2 = 0.025$ probability for 49 degrees of freedom would be selected from Table B.2 as 2.01 (noting that where the value does not appear in the table it is necessary to take the lowest closest value). In this case the calculated value is also outside the critical region and so we reject the null hypothesis. Our conclusion is that, at the 0.05 significance level, there is a difference between the sampled magnesium concentrations and their stated value. Note for comparison that the Z-test gave a critical value of 1.96 and so if a Z-test is not justified here then whilst a t-test may be appropriate the difference must be slightly larger in order that we may declare it significant.

The p-value for the t-test can be calculated using Excel's function =tdist(2.32, 49, 2). Note that Excel requires the absolute value of t_{calc} (i.e. for the minus sign to be ignored), and also allows you to specify a two-tailed test which results in a p-value of 0.025. R can compute the p-value too, using the function pt(-2.32, 49) but note that R requires the minus sign (and if the test statistic is positive, requires the user to subtract the result from 1) and provides the one-tailed probability only (in this case 0.012 – you must multiply by two if you are doing a two-tailed test).

5.3 Comparing proportions for one sample

Quite often the subject of interest will be the fraction of a particular population that falls into one category rather than another. A typical example of this situation would be if a researcher were interested in deciding whether the proportion of men versus women in a population was a certain value. The mathematics of this situation is slightly different from the cases encountered previously, but fortunately when the sample size is large and the distribution between the groups is not too extreme, it is possible to use the normal distribution to test the hypothesis that the population proportion is equal to a known value, P_0. The appropriate test statistic is:

$$Z_{\text{calc}} = \frac{X - nP_0}{\sqrt{nP_0\left(1 - P_0\right)}}, \qquad (5.5)$$

which follows a normal distribution, providing that $nP > 5$ and $n(P - 1) > 5$. The critical region for this test can be defined such that in the two-tailed case where a difference from P_0 in either direction is of equal interest we reject H_0 when $Z_{\text{calc}} > z_{\alpha/2}$ or $Z_{\text{calc}} < -z_{\alpha/2}$.

Worked Example 5.4 Suppose one were interested in knowing whether the number of women employed by a firm, X, represented a proportion different from 50%, that is $H_0: P = 0.5$; $H_1:$ $P \neq 0.5$, where $P = X/n$. If a random sample of 100 members of staff gave $X = 40$ determine whether the proportion of men and women is significantly different from 50%.

Solution

In the present example, $Z_{calc} = \dfrac{40 - (100)(0.5)}{\sqrt{100(0.5)(1-0.5)}} = \dfrac{-10}{5} = -2.0.$

Noting from Table B.1 that $z_{\alpha/2} = -1.96$ and that Z_{calc} is less than this value, we reject the null hypothesis and accept the alternative hypothesis, to conclude that women do not represent 50% of the people employed by this company. We can of course calculate a p-value for this test, using the methods described in Section 5.1.2. The resulting p-value is 0.023, which is less than $\alpha/2 = 0.025$, but only just.

5.4 Comparing two samples

5.4.1 Independent samples

Whilst earlier examples have been concerned with comparing the mean of one sample against a given value, where the given value is known without error, it is often necessary to compare the mean of two samples where there is the possibility of error in each estimate. This approach often arises in comparative studies in which the researcher is interested in distinguishing between means of two independent samples that are not related to each other. Various tests can be used to address this type of comparison between independent samples, including some which use the standard normal distribution for large samples with known variances and others which use the t-distribution for small sample sizes with unknown variances. It is rare in geographical cases to have enough data for a large sample test, and even rarer to know the variances *a priori*, so only the procedures based on the t-distribution for use with samples of unknown variance are presented here.

5.4.2 Comparing means: *t*-test with unknown population variances assumed equal

The parameter of interest in this situation is the difference between the means, μ_1 and μ_2, of two normally distributed variables whose variances are unknown but which are assumed to be equal. The basic procedure is similar to the one-sample t-test above. The null hypothesis is:

$$H_0 : \mu_1 - \mu_2 = d,$$

and the alternative hypothesis is:

$$H_1 : \mu_1 - \mu_2 \neq d.$$

Here d is the hypothesized difference between the two means, and is often chosen to be zero in cases where the null hypothesis covers the situation that there is no difference between the two means. Just as in the earlier case where it was sensible to estimate the unknown population variance using the sample variance, here it is appropriate to combine the sample variances to give an estimate of the pooled variance, S_P^2, which is defined as:

$$S_P^2 = \frac{(n_1 - 1)S_1^2 + (n_2 - 1)S_2^2}{n_1 + n_2 - 2}, \tag{5.6}$$

where S_1^2 and S_2^2 are the variances of the two samples in question. A closer look at Equation 5.6 reveals that although it looks a bit complicated it is in fact a weighted average of the two separate sample variances. Using the pooled variance, the test statistic is then calculated in a very similar way to before, to give:

$$t_{calc} = \frac{\overline{X}_1 - \overline{X}_2 - d}{S_P \sqrt{\dfrac{1}{n_1} + \dfrac{1}{n_2}}}, \tag{5.7}$$

which follows a t-distribution with $n_1 + n_2 - 2$ degrees of freedom. It is worth noting that the number of degrees of freedom is the total number of observations minus two to cover the two parameters (the two means) that have been calculated. The number of unrestricted observations is therefore $n_1 + n_2 - 2$.

Worked Example 5.5 An environmental scientist wishes to compare faecal coliform (bacteria) levels on two beaches. Based on a random sample of measurements taken from waters at each beach, reported in Table 5.1, determine with 0.05 significance whether the two beaches are equally polluted. Use a t-test, together with the information contained in the table to test the null hypothesis that the beaches have the same level of faecal coliform contamination. State any assumptions that you make.

Table 5.1 Beach pollution.

Beach #1 (colonies/100 ml)	Beach #2 (colonies/100 ml)
118	101
104	130
92	56
112	112
169	127
160	24
85	88
140	110
107	104
165	86
75	69
123	77
104	33
169	92

Solution The first step is to calculate summary statistics for each dataset. These are the mean, variance, and standard deviation:

	Beach #1	Beach #2
Count	14	14
Mean	123.1	86.4
Variance	1037.5	1029.2
Std dev.	32.2	32.1

Note that the mean coliform concentration is almost 50% higher for Beach #1, but that the two samples have very similar variances. To compare the mean coliform concentrations, a pooled *t*-test is suitable. The question asks if the concentrations are different (rather than specifying a direction for the difference), so a two-tailed test is required with H_0: $\mu_1 = \mu_2$ and H_1: $\mu_1 \neq \mu_2$. A suitable significance level, $\alpha = 0.05$ (i.e. 95% confidence interval). In this case, using Equation 5.6 gives a pooled variance of 1035.5 and Equation 5.7 gives $t_{calc} = 3.02$, and the critical value of the t-distribution, $t_{crit} = 2.06$ (using $\alpha = 0.05$, with $14 + 14 - 2 = 26$ degrees of freedom). In Excel use =tinv(0.05,26), noting that Excel automatically divides α by 2 for a two-tailed test. Since $t_{calc} > t_{crit}$, the null hypothesis can be rejected. That is, at the 0.05 significance level, we can conclude that there is a significant difference between the concentrations of faecal coliforms present on the two beaches. A quick glance at the mean values shows that Beach #2 has the lower of the two levels.

Note that both Excel and R provide built-in functions for performing t-tests. These do not eliminate the need to understand the basic assumptions behind the tests but they do reduce the tedium of the calculations. Excel's function for performing two-sample t-tests is called =t.test() and it requires the user to specify the two arrays containing the two samples along with the number of tails and the type of test (see Excel's built-in help for the types of test available). In return, Excel gives the p-value associated with the test. For the example above, the value returned is 0.0056, which is much lower than 0.05 and indicates that the difference between the faecal coliform concentrations at the two locations is significantly different from zero.

The same procedure can be followed in R, using the commands:

```
coliforms <- read.csv("wex5_1.csv")

t.test(coliforms$Beach1,coliforms$Beach2,
    mu=0, alternative="two.sided", var.
    equal=TRUE)
```

The first line reads in the data, and the second line performs the test. The user can specify the difference that is hypothesized (referred to as mu by R, which corresponds to d in the discussion above and is often zero). It is also possible to select the correct alternative hypothesis (in this case "two.sided" specifies a two-tailed test), and to inform R that we wish to assume that the variances are equal. The resulting output is:

```
            Two Sample t-test

data: coliforms$Beach1 and coliforms$Beach2

t = 3.0218, df = 26, p-value = 0.005581

alternative hypothesis: true difference in
   means is not equal to 0

95 percent confidence interval:

   11.74014 61.68843

sample estimates:

mean of x mean of y

123.07143 86.35714
```

In which it should be noted that R has produced values of the *t*-statistic and the *p*-value which are in agreement with those calculated by hand above. R goes further and reports a confidence interval based on the *t*-distribution for the difference between the two means, and gives the sample estimates of the means themselves.

5.4.3 Comparing means: *t*-test with unknown population variances assumed unequal

In situations where the (unknown) population variances are not approximately equal, an alternative form of the *t*-test can be used, called Welch's test. Obviously it is not possible to be certain whether the population variances are equal if they are not known, but the equality of the sample variances gives a good indication. A discussion of some more precise ways to test for equality of variances is given in Section 5.4.5.

The formulation of the null hypothesis is identical, so H_0: $\mu_1 - \mu_2 = d$, as above. The test statistic is actually simpler to calculate.

$$t_{calc} = \frac{\bar{x}_1 - \bar{x}_2 - d}{\sqrt{\dfrac{S_1^2}{n_1} + \dfrac{S_2^2}{n_2}}}, \tag{5.8}$$

but the calculation of the number of degrees of freedom, denoted ν (a Greek letter 'nu'), is a little more tedious:

$$v = \frac{\left(\dfrac{S_1^2}{n_1} + \dfrac{S_2^2}{n_2}\right)^2}{\dfrac{\left(S_1^2 / n_1\right)^2}{n_1 - 1} + \dfrac{\left(S_2^2 / n_2\right)^2}{n_2 - 1}}. \tag{5.9}$$

This procedure sometimes gives a non-integer number of degrees of freedom, which should be rounded down to the next integer value for use with statistical tables. With the null hypothesis, the test statistic and the number of degrees of freedom it is possible to determine whether to reject or accept the null hypothesis, assuming that the test statistic takes a t-distribution with v degrees of freedom at the $\alpha/2$ significance level, or to calculate a p-value using the techniques described in Section 5.2.

Worked Example 5.6   A climatologist is interested in the differences in rainfall between Glasgow and London given in Table 2.5. Using a t-test, test the hypothesis that there was no difference between the average rainfall amounts in the two locations during the period 1981–2010, against the alternative that there was a difference. Use a 0.05 level of significance.

Solution   A t-test is necessary to compare the means here, but the summary statistics in Table 5.2 show that the variances differ by a factor of greater than three so we should not assume that they are equal. This means we need a t-test for unequal variances. The null and alternate hypotheses are H_0: $\mu_1 = \mu_2$ and H_1: $\mu_1 \neq \mu_2$, so the test is two-tailed, because the direction of the difference is not specified beforehand, and a significance level of 0.05 is suitable.

The test statistic is calculated using Equation 5.8, which gives $t_{\text{calc}} = 17.0$, and the number of degrees of freedom calculated from Equation 5.9 is 45. This gives $t_{\text{crit}} = 2.32$ ($\alpha = 0.05/2$, and $v = 45$). In this example, t_{calc} is greater than t_{crit}, so we reject H_0 and state that there is evidence to conclude that the mean rainfall in Glasgow is different from that in London. Then, looking at the respective means it can be stated that, with 95% confidence, rainfall is higher in Glasgow than in London.

Table 5.2 Summary of rainfall data in Glasgow and London.

	Glasgow	London
Count	30	30
Sum	37353	18048
Mean	1245	602
Var	32740	10264

This test can be performed with Excel with the same approach used in the previous example, except that it is necessary to select the option for unequal variances. The reported p-value is 2.45E-21 or 2.45×10^{-21}, which is a very small number indeed.

In R, the approach is similar too, but the analyst must state that var.equal = FALSE in the command line:

```
rainfall <- read.csv("wex5_6.csv")

t.test(rainfall$Glasgow,rainfall$London,
  mu=0, alternative="two.sided", var.
  equal=FALSE)
```

Again the results are identical (although R simply states that the p-value is lower than 2.2×10^{-16}, which is the lowest number that the software can reliably handle without potentially suffering from rounding errors):

```
    Welch Two Sample t-test

data: rainfall$Glasgow and rainfall$London

t = 16.9913, df = 45.543, p-value < 2.2e-16

alternative hypothesis: true difference in
means is not equal to 0

95 percent confidence interval:

  567.2464 719.7536

sample estimates:

mean of x mean of y

1245.1333 601.6333
```

5.4.4 t-test for use with paired samples (paired t-test)

An experiment or monitoring programme can be particularly effective where samples can be taken in pairs. Each pair of observations is made under similar conditions, but these conditions may change from pair to pair. An example would be repeated measurements of the same quantity at two time periods, perhaps if a hydrologist were interested in monitoring the acidity of a population of lakes before and after a new law was introduced to reduce pollution. In these circumstances, a good research strategy would be to take measurements in each of the lakes before and after the new law came into effect. One approach would be to investigate whether the average pH in the lakes had changed, by using the independent samples t-test for differences in the mean before and after the pollution stopped. This would be a valid approach, but it does not take advantage of the fact that the measurements are paired: that is, they are from the same lakes. It would be good to take advantage of the fact that conditions in each lake have (we expect) remained constant except for the change that is of interest. The procedure for taking advantage of the paired measurements is to use a paired t-test instead. Here, the aim is to look at the difference between the two measurements in each lake and see if the *differences* are significantly different from zero using a one-sample t-test. The benefit of this approach is we are only ever comparing like with like because the pairing process removes all of the variance between lakes. The variance that appears in the t-test is therefore only concerned with differences between the two time periods. Providing that the pairs are chosen carefully, this is a much better experimental design. The null hypothesis is H_0: $\mu_D = d$, and the test statistic is:

$$t_{calc} = \frac{\bar{D} - d}{S_{\bar{D}} / \sqrt{n}}, \qquad (5.10)$$

where $\bar{D}$ is the sample mean of the differences between paired observations, $S_{\bar{D}}$ is the standard deviation of the differences, and d is the hypothesized difference (usually zero if the null hypothesis is that there is no difference between the values in each pair). The test statistic follows a t-distribution and so the null hypothesis is rejected if $t_{calc} > t_{\alpha/2, n-1}$ or $t_{calc} < -t_{\alpha/2, n-1}$.

Worked Example 5.7   A government agency wishes to measure the effectiveness of a policy designed to reduce pollution due to acid rain by monitoring the pH (acidity) of 22 lakes. Paired samples were taken from each lake: one sample from before the policy was introduced, the other some 10 years afterwards (Table 5.3). Use a paired *t*-test to evaluate the effectiveness of this pollution reduction scheme.

Table 5.3 Comparison of lake acidity before and after pollution reduction programme.

	pH	
Lake	Before	After
A	6.39	6.24
B	6.45	6.43
C	5.81	5.84
D	5.41	5.61
E	5.46	6.16
F	6.11	6.21
G	4.90	5.44
H	4.64	4.95
I	5.44	6.02
J	4.99	5.42
K	6.50	6.70
L	5.44	6.35
M	4.55	5.23
N	5.71	5.87
O	5.23	5.83
P	5.35	7.68
Q	5.29	5.92
R	5.51	6.10
S	5.78	5.95
T	5.19	5.26
U	4.66	5.07
V	6.56	6.20

Solution These observations are paired, which means that each pair of observations was taken in similar conditions, but that these conditions may change from one pair to another (in this case samples are taken from the same lake). The required test is a paired t-test in which the differences between the lake pH values for each pair is tested against the null hypothesis that the difference is zero. The analyst is concerned with detecting an increase or a decrease in lake pH, and so a two-tailed test, with 0.05 significance level is appropriate. The test statistic, $t_{calc} = 2.97$ and $t_{crit} = -2.08$ ($\alpha = 0.05/2$, with $n - 1 = 12$ degrees of freedom). Note that $|t_{calc}| > |t_{crit}|$, so the null hypothesis is rejected and we conclude that there is evidence for a change in lake pH. The change is positive and indicates that pH increased by 0.41 (i.e. acidity decreased) between the two time periods.

Excel's $= T.TEST()$ function can be used here too (by using option 1 for type). It returns a p-value of 0.0015, which is consistent with the conclusion reached above. Note that if the standard t-test is performed on these data instead of the paired t-test the p-value is 0.02807, which is much higher. This demonstrates the value of the paired experimental design.

Paired t-tests can be performed in R by setting the keyword 'paired = TRUE' in the command line:

```
lakes <- read.csv("wex5_7.csv")

t.test(lakes$Before,lakes$After, mu=0,
    alternative="two.sided", paired=TRUE)
```

The resulting output gives identical results to those obtained by hand and with Excel:

```
        Paired t-test

data:   lakes$Before and lakes$After

t = -3.6575, df = 21, p-value = 0.001469

alternative hypothesis: true difference in
    means is not equal to 0

95 percent confidence interval:
  -0.6495380 -0.1786438

sample estimates:
mean of the differences
        -0.4140909
```

5.4.5 Comparing variances: *F*-test

So far the tests presented here have considered comparisons between the *means* of different samples. It can also be useful to compare the *variances*, especially in cases where a change in the variability of a phenomenon is as interesting as a change in its mean. Another application in which the comparison of variances is often useful is in checking the assumptions that are prerequisites for further analysis have been met. For example, in the earlier discussion of the *t*-test in Section 5.4.2, one of the necessary assumptions was that the variances were equal. The comparison of variances also forms the basis for a very popular and powerful series of statistical techniques called 'analysis of variance' or ANOVA which is described in more detail in Chapter 7.

The *F*-test compares two variances. The ratio of two variances takes the *F*-distribution and so to test the null hypothesis, H_0: $\sigma_1^2 = \sigma_1^2$, the *F*-statistic is calculated:

$$F_{\text{calc}} = \frac{S_1^2}{S_2^2}, \qquad (5.11)$$

where S_1^2 and S_2^2 are the two sample variances in question and where S_1 is the larger of the two variances in question. The *F*-distribution is given in Table B.3 and can be computed using Excel or R. To calculate the size of the critical region using the *F*-distribution requires the number of degrees of freedom associated with each of the variances, such that in the two-tailed case the null hypothesis is rejected if $F_{\text{crit}} > f_{\alpha/2, n_1-1, n_2-1}$ or $F_{\text{crit}} < -f_{\alpha/2, n_1-1, n_2-1}$.

Worked Example 5.8 Use an *F*-test to decide whether variances of the rainfall data from London and Glasgow (Table 2.5) are equal.

Solution The test statistic, $F_{\text{stat}} = (32\ 740)/(10\ 264) = 3.19$. Note that it is important to put the larger of the two variances as the numerator here. Thus F_{stat} is greater than $F_{\text{crit}} = 1.88$, for $\alpha = 0.05$ with 28 and 28 degrees of freedom, which can be obtained in Excel as =fdist(0.05,28,28) or in R with pf(0.975,28,28) =1.88 (noting that in R it is necessary to give the probability $p = 1 - \alpha/2 = 0.975$). It is also worthwhile to note that here we are performing a two-tailed test. We reject the null hypothesis and conclude that the variances are not equal at the 0.05 significance level. This conclusion directs us to use Welch's test rather than the standard t-test, which requires equal variances.

The theory behind the F-test relies on the underlying data being normally distributed. In practice this means that if the F-test indicates that variances are unequal it may simply be that they do not follow a normal distribution. In such cases, there are other tests for equality of variances whose assumptions are less stringent than those underlying the F-test. The most common is Bartlett's test, which is explained in Section 7.2.

5.5 Non-parametric hypothesis testing

5.5.1 Parametric and non-parametric tests

All of the tests encountered so far have assumed that the data come from a particular probability distribution, which can be characterized by its parameters (e.g. the mean and standard deviation). In this case the assumption has been that the data take the form of a normal distribution. In practice this requirement may often not hold, and a set of more general methods is required that do not assume a theoretical distribution according to which the data are distributed. Such methods are referred to as *distribution free* and in many cases they involve making calculations that do not involve the estimation of parameters such as the mean or variance, hence they may also be termed *non-parametric*.

In practice it is usually best to employ a parametric method where the extra assumptions are met because non-parametric tests tend to be less 'powerful', which is to say that they require the presence of a greater level of difference before that difference is judged to be statistically significant. Many non-parametric methods rely on order statistics (i.e. ranked data) and so they can often be applied to data which were collected only at the ordinal level in the first place, such as survey responses.

5.5.2 Mann–whitney *U*-test

When the underlying dataset is not normally distributed, the t-test cannot be used. In such circumstances a non-parametric test is required. The most commonly used non-parametric test for deciding whether there is a difference between the means of two independent samples is the Mann–Whitney U-test. This test is useful because: (i) it does not assume that data are normally distributed; (ii) it can be used with ordinal (ranked) or interval (measured) data; and (iii) it is appropriate where sample sizes are small. This test is sometimes also referred to as the Wilcoxon rank sum test. The null hypothesis is that the two datasets are drawn from the same population, and the alternative hypothesis is that they are drawn from different populations. This alternative hypothesis results in a two-tailed test. The data are first ranked without taking account of which sample the observation is in. If there are any ties, then the tied values are given the mean of the ranks that they would have had. The test statistic, U, is calculated as the lesser of U_1 and U_2, which are defined as:

$$U_1 = n_1 n_2 + \frac{n_1(n_1 + 1)}{2} - \sum r_1, \text{ and} \tag{5.12}$$

$$U_2 = n_1 n_2 + \frac{n_2(n_2 + 1)}{2} - \sum r_2, \tag{5.13}$$

where n_1 and n_2 are the number of observations in each sample, and r_1 and r_2 are the computed ranks.

The Mann–Whitney U-test is non-parametric, which brings a significant advantage in many geographical applications where datasets often do not follow a normal distribution. Moreover, because this test works with ranks, it can be used for data that are measured on an ordinal scale, that is, data that were originally collected as ranks. The test is therefore more widely applicable than the t-test or Z-test, and yet requires only a limited extra amount of calculation. It should be noted that if the assumption of normality does hold, the Mann–Whitney U-test will still produce useable results, but the p-value would have been slightly lower had a t-test been used instead (assuming all its assumptions had been met).

Worked Example 5.9 A geography student is interested in measuring the importance attached to job security by employees in two different companies. The student interviews ten

Table 5.4 Job security data.

Firm A	Firm B
4	2
1	2
3	5
3	4
1	2
2	5
3	5
2	4
1	3
2	2

randomly chosen employees from each firm and asks them to express the importance of job security to them on a scale of 1 to 5, giving the data in Table 5.4. Use a Mann–Whitney U-test to determine whether, at the 0.05 significance level, the two samples are drawn from the same population.

Solution The first step in this analysis is to rank each data point relative to the dataset as a whole without regard for which sample each data point belongs to. Tied ranks must be assigned the average of the ranks that they would have been given (Table 5.5).

In the present example, $n_1 = n_2 = 10$, $\sum r_1 = 80.5$ and $\sum r_2 = 129.5$ and so $U_1 = 10^2 + (110/2) - 80.5 = 74.5$, and $U_2 = 10^2 + (110/2) - 129.5 = 25.5$, meaning U is assigned the lower of the two, $U = 25.5$. The critical value for U is given as 23 in Table B.4 (with 0.05 significance and the appropriate sample sizes) and so with the calculated value being greater than the critical value, the null hypothesis cannot be rejected and we cannot claim that the two samples come from different populations.

Unfortunately, Excel does not have a built-in function to calculate the Mann–Whitney U-test and, although it is possible to perform the calculations by hand with Excel, it is inadvisable to rely on Excel's built-in ranking function because it does not

Table 5.5 Ranked job security data.

Firm A	Firm B
16.0	7.0
2.0	7.0
12.5	19.0
12.5	16.0
2.0	7.0
7.0	19.0
12.5	19.0
7.0	16.0
2.0	12.5
7.0	7.0

average tied ranks. However, it is possible to perform this test with R using the Wilcoxon rank sum test, `wilcox.test()`, which is identical to the Mann–Whitney U-test:

```
jobsec <- read.csv("wex5_9.csv")
```

```
wilcox.test(jobsec$A,jobsec$B)
```

This produces the following output:

```
      Wilcoxon rank sum test with continuity
         correction
```

```
data:  jobsec$A and jobsec$B
```

```
W = 25.5, p-value = 0.0615
```

```
alternative hypothesis: true location shift
   is not equal to 0
```

```
Warning message:
```

```
In wilcox.test.default(jobsec$A, jobsec$B) :
   cannot compute exact p-value with ties
```

from which it can be seen that the test statistic is the same (in spite of its being referred to as W rather than U), and that the p-value is 0.06 which is consistent with our failure to reject the null hypothesis at the 0.05 significance level.

Note that, as explained in the output message, in the presence of tied ranks it is not possible to calculate precise p-values for this test, but the approximation used by R, which is enhanced by incorporating a continuity correction to the calculation of the p-value, is satisfactory for most practical purposes. Note that R's description of the alternative hypothesis as 'location shift is not equal to 0' is equivalent to stating that there is no systematic difference between the ranks of the two datasets.

Exercises

1 Compare the assumptions which are made in the: (i) Z-test; (ii) t-test; and (iii) Mann–Whitney U-test.

2 Use a t-test to evaluate whether there is a difference between the lead concentration measurements for the two sites reported in Table 2.3.

3 Repeat Worked Example 5.7 using a t-test for independent samples instead of a paired t-test. What implications do your findings have for research design?

6

Comparing distributions: the Chi-squared test

STUDY OBJECTIVES

- Understand the issues involved in the comparison of distributions.
- Know how to construct a contingency table and populate it with observed and expected frequencies.
- Be able to calculate the Chi-squared test statistic based on data reported in a contingency table and to interpret its meaning with reference to critical values of the Chi-squared distribution.
- Be aware of the assumptions on which the Chi-squared statistic is based and the resulting restrictions on the test's validity.

6.1 Chi-squared test with one sample

At the heart of many geographical studies lies the question of whether two datasets are drawn from the same statistical distribution. For example, does the distribution of exam achievement vary with gender or socioeconomic class, or does the observed distribution of plant types along a transect differs from that which would be predicted by the latest theory? The one-sample Chi-squared test (often written χ^2) offers a very flexible approach to address this type of question. It is a technique which is particularly flexible because it uses only counts or frequency data and is non-parametric, which means that there is no underlying assumption of normality.

Statistical Analysis of Geographical Data: An Introduction, First Edition.
Simon J. Dadson.
© 2017 John Wiley & Sons Ltd. Published 2017 by John Wiley & Sons Ltd.

The first step involved in a Chi-squared test is the construction of a contingency table. This is a table in which all of the possible observed outcomes are recorded as counts or frequencies. These observations are then compared with the theoretical expectation of outcomes. The null hypothesis in such a test in that the observed frequencies are consistent with the expected frequencies. In the one-sample Chi-squared test, unless there is a good reason to do otherwise, it is usual to calculate the expected frequencies using a uniform distribution. The expected frequencies are assumed to be identical and are calculated simply by dividing the total number of observations by the number of categories used.

Once the observed and expected frequencies are known, the Chi-squared statistic can be calculated as:

$$\chi^2 = \sum \frac{(O_i - E_i)^2}{E_i}, \tag{6.1}$$

where O_i and E_i are the observed and expected counts, respectively. The test statistic follows a Chi-squared distribution with the number of degrees of freedom equal to the number of categories minus one (see Appendix B, Table B.5).

A number of prerequisites are necessary in order to obtain meaningful results with the Chi-squared test. The first is that the data in each category must be counts, not proportions or percentages. The second requirement is that no category should have an expected frequency less than five. Some statisticians relax this threshold to one or two when the number of categories is large providing that more than 80% have an expected frequency above five (Roscoe and Byars, 1971). In practice this condition can usually be met by aggregating categories, providing that this does not compromise the nature of the study. Moreover, it should be noted that whilst this test can detect differences between the observed and expected distributions, it does not give any information on what the differences are.

Worked Example 6.1 A geographer is interested in finding out whether visitors to a large museum in their home town are drawn uniformly from across a range of age groups or whether certain age groups are more or less likely to visit. The geographer

Table 6.1 Museum visitor ages.

	<16	16–30	31–45	46–60	>60
No. of visitors	18	8	14	28	32

Table 6.2 Contingency table for museum visitor ages.

	<16	16–30	31–45	46–60	>60
O_i	18	8	14	28	32
E_i	20	20	20	20	20
O_i-E_i	−2	−12	−6	8	12
$(O_i-E_i)^2$	4	144	36	64	144
$(O_i-E_i)^2/E_i$	0.2	7.2	1.8	3.2	7.2

O, observed; E, expected.

states the null hypothesis to be that visitor numbers are uniform by age group, with the alternative hypothesis being that the distribution is not uniform. The geographer then asks a random sample of 100 visitors and finds the results given in Table 6.1. Do these values differ from what we might have expected under the null hypothesis?

Solution Under the null hypothesis, we would have expected an equal number of visitors (100/5 = 20) in each of the five age groups, so the contingency table would be as shown in Table 6.2.

To calculate the Chi-squared statistic, the values of $(O_i-E_i)^2/E_i$ in the bottom row of Table 6.2 are simply added up to give $\chi^2 = 19.6$. Intuitively, it can be seen that if there are large differences between the observed counts and the expected counts the Chi-squared statistic will be higher than if the O_i and E_i are similar. For example, if the number of observations in each class were the same as expected (i.e. 20) then the test statistic would equal zero indicating no difference from the expected distribution. To test whether the departure from the expected distribution is statistically significant the number of degrees of freedom must be calculated. It is one fewer than the number of categories, so $5 - 1 = 4$. The critical value of the Chi-squared distribution is therefore 9.49 (at the 0.05 significance level) and so because the calculated value exceeds the critical value we reject the null hypothesis and conclude that there is a distribution of museum attendance is not uniform across age group. Alternatively, we can compute the *p*-value using Excel's =CHIDIST (19.6, 4) function which returns a value of 0.0006 and leads to the same conclusion.

The Chi-squared test can be performed in its entirety in Excel using a function called =CHITEST (), which returns the *p*-value of a one-sample Chi-squared test given the observed and expected frequencies as arguments. The equivalent procedure in R uses the built-in function chisq.test ():

```
museum <- c(18,8,14,28,32) #simple datasets
  can be entered manually
chisq.test(museum)
```

and produces both the test statistic and the *p*-value as output:

```
    Chi-squared test for given probabilities
data: c(18, 8, 14, 28, 32)
X-squared = 19.6, df = 4, p-value = 0.0005989
```

6.2   Chi-squared test for two samples

The Chi-squared test can also be used with two samples so that the distribution of observed counts among different categories can be compared. The null hypothesis in the two-sample case is that the two samples are drawn from the same distribution; the formula used to calculate the test statistic is identical to the one-sample version of the test; and the calculation of expected frequencies is only slightly more involved because they must be calculated based on the row and column totals. First, the row and column totals are calculated and then the expected values are calculated for each cell in the table as the product of the corresponding row and column total divided by the overall table total. For each cell, the squares of the differences between the observed and expected values are divided by the expected

Table 6.3 River habitat biodiversity.

	High	Good	Moderate	Poor	Bad
Farmland	14	24	20	15	27
Woodland	42	25	19	15	13

value, according to Equation 6.1. The sum of these numbers gives the Chi-squared statistic. In the two-sample case, the number of degrees of freedom is given as:

$$v = (r-1)(c-1). \tag{6.2}$$

Worked Example 6.2 Suppose that an ecologist is interested in stream biodiversity and would like to find out whether biodiversity is different in areas of farmland compared with areas of woodland. The ecologist takes a random sample of river reaches in different land-cover types and classifies each river reach into one of five groups based on a predetermined set of biodiversity criteria. The results are given in Table 6.3.

Use a two-sample Chi-squared test to test the null hypothesis that there is no difference between the distributions against the alternative that there is such a difference. Use a 0.05 significance level.

Solution The first step is to take the contingency table and compute the row and column totals (Table 6.4). These are then used to calculate the expected frequencies for each cell as the product of the relevant row and column total divided by the overall total (Table 6.5). For example, for Farmland with a high biodiversity index the expected frequency is found by multiplying the row total for 'Farmland', which is 14 by the column total for 'High' which is 56 and dividing by the overall total which is 114, to give $(14 \times 56)/114 = 26.3$, and so on.

Once the expected frequencies have been calculated the next step is to calculate the differences between observed and expected frequencies and divide them by the expected frequency (see Equation 6.1). These quantities are then added to find the calculated value of the Chi-squared statistic, which in

Table 6.4 Contingency table for stream habitat biodiversity showing observed counts and calculated row and column totals.

	High	Good	Moderate	Poor	Bad	Total
Farmland	14	24	20	15	27	100
Woodland	42	25	19	15	13	114
Total	56	49	39	30	40	214

Table 6.5 Contingency table for stream habitat biodiversity showing expected frequencies.

	High	Good	Moderate	Poor	Bad
Farmland	26.2	22.9	18.2	14.0	18.7
Woodland	29.8	26.1	20.8	16.0	21.3

this case is 18.1. The number of degrees of freedom associated with this statistic is given by Equation 6.2 as $(2-1)(5-1) = 4$. The corresponding critical value of Chi-squared is found to be 9.5 from Table B.5 by selecting probability = 0.05 and four degrees of freedom. Thus the null hypothesis is rejected and it is concluded that there is a significant difference between the two distributions.

Excel's =CHITEST() function can cope with a two-sample Chi-squared test simply by providing it with the table of observed and expected frequencies. It calculates, but does not report, the test statistic and returns the p-value for the test, in this case 0.001. R provides more meaningful output when the two samples are given to chisq.test():

```
biodiversity <- read.csv("wex6_2.csv",
   header=TRUE, row.names=1)

chisq.test(biodiversity)
```

Note that it is necessary to tell R that row names have been supplied otherwise it tries to interpret the row names as data values. The results are identical to those of the manual calculation:

```
Pearson's Chi-squared test

data: biodiversity
X-squared = 18.1077, df = 4, p-value =
   0.001176
```

Exercises

1 A geographer is interested in the occurrence of graffiti on public buildings and hypothesizes that the type of public building strongly influences the likelihood that it will attract graffiti. The geographer collects data by examining ten buildings chosen at random from each of the four categories listed in Table 6.6. Analyse the data to determine whether there is such an association. Explain the assumptions behind the Chi-squared test and the extent to which they are met.

Table 6.6 Distribution of graffiti by building type.

	Shop	Public lavatory	School	Church
No. of premises with graffiti present	8	10	5	1

2 An ecologist is interested in the effect of natural disturbance on the availability of seeds for birds. The numbers of small, medium and large seeds sampled before and after a major storm are given in Table 6.7. Use a Chi-squared test to evaluate the hypothesis that there is a change in the distribution of available seed sizes. Select an appropriate significance level, and state any assumptions that you make.

Table 6.7 Distribution of bird seed availability.

	Small	Medium	Large
Before	11	26	37
After	28	24	16

3 An education specialist is interested in the beneficial effects of preschool education on literacy and collects the data shown in Table 6.8 on the improvement in literacy scores between ages 3 and 5 amongst children who do or do not attend preschool. Use a Chi-squared test to evaluate the

Table 6.8 Effects of preschool education on literacy.

	Much worse	Worse	Better	Much better
No preschool	10	32	43	15
Preschool	8	18	49	25

hypothesis that preschool education improves children's literacy. Comment on your results. How might the study have been improved to provide more informative conclusions?

7

Analysis of variance

STUDY OBJECTIVES

- Understand why analysis of variance (ANOVA) is used to compare datasets when more than two samples are involved.
- Understand the calculation for total, treatment, and error sums-of-squares and the associated partitioning of variances between the sources of variation.
- Be able to calculate the appropriate number of degrees of freedom for each term in the ANOVA.
- Know how to construct an ANOVA table and how to use it to calculate mean-squares for treatment and error terms and to use these to compute the resulting value of F_{calc}.
- Interpret the calculated value of F_{calc} either by using the appropriate critical value found from a table or by calculation of a p-value.
- Have an awareness of assumptions involved in ANOVA and know techniques available to diagnose departures from those assumptions.
- Perform appropriate multiple comparison tests after ANOVA with an awareness of the effect on statistical significance.
- Recognize occasions when standard ANOVA is not appropriate and in those cases to be able to use non-parametric alternatives.

Statistical Analysis of Geographical Data: An Introduction, First Edition.
Simon J. Dadson.
© 2017 John Wiley & Sons Ltd. Published 2017 by John Wiley & Sons Ltd.

7.1   One-way analysis of variance

Comparing the means of two samples is straightforward using a Z-test or a t-test. The aim of this chapter is to generalize the question of comparison of means to cater for three or more samples. Whilst it would be possible to compare more than two samples using a combination of t-tests, doing so would be difficult because the number of t-tests that would be needed would rapidly become unmanageable. With five samples, for example, 10 t-tests would be needed. With 10 samples, 45 t-tests would be needed to compare each of the possible pairs of samples. An additional (and more important) problem is that with each of these t-tests there comes a 5% probability of obtaining a significant difference by chance (i.e. of making a Type I error, see Section 3.2). If more than one t-test is carried out, then the probability of making a Type I error is compounded and there is an increased chance of concluding that there is an effect when there is in fact not. The significance level associated with multiple comparisons is $(1 - \alpha)^n$ where n is the number of tests and α is the significance level used in each individual test. So for pair-wise multiple comparisons involving 10 samples with each test performed at $\alpha = 0.05$, the new level of significance is $\alpha = (0.05)^{10} = 0.40$. In other words, rather than maintaining a 5% likelihood of finding with the pair-wise comparison there is now a 40% chance of finding a significant result by chance.

Instead of comparing the means directly, a better approach is to determine whether the variation between the samples is large relative to the variation within them. This approach is called analysis of variance, or ANOVA. It is widespread and it proves to be a very useful strategy for data analysis especially in designed experiments. The best illustration of how ANOVA works is through an example.

Suppose that a farmer is interested in increasing crop yields and wishes to investigate which of three different crop varieties gives the greatest yield. The farmer has 30 plots available for the experiment and allocates each plot a crop type at random according to the diagram in Figure 7.1. This is a completely randomized experimental design with one factor (crop variety). There are three levels (or treatments: Crop A, Crop B, and Crop C), each of which has 10 replicates.

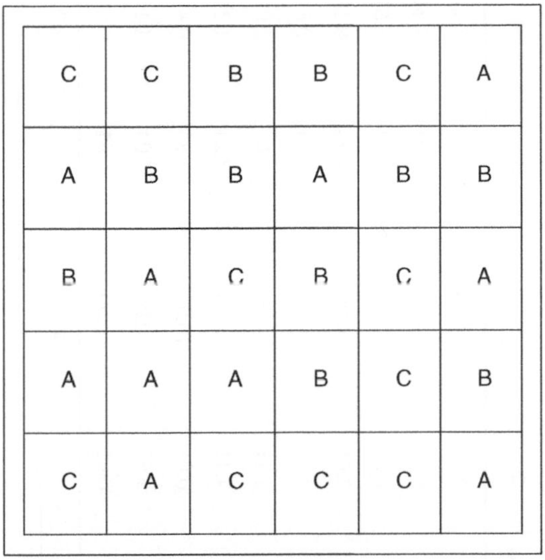

C	C	B	B	C	A
A	B	B	A	B	B
B	A	C	B	C	A
A	A	A	B	C	B
C	A	C	C	C	A

Figure 7.1 Completely randomized experimental design.

One obvious problem that the farmer needs to take into account is that the plots themselves may not be identical: some may be more or less fertile than the others, or wetter, or better drained. If one crop is preferentially allocated to the best plots, then any difference between the effect of crop type will be mixed up with the differences in the properties of the plots themselves. The randomization is therefore an essential feature of the experimental design. By assigning the crops to the fields at random, the farmer will average out any variability that might arise through differing fertility of the fields, with the aim that only effects which are due to the different treatments should appear systematically in the results of the study. When it comes to analysing the results, the farmer can then ask which is bigger: the variability in crop yields *within* the set of plots that were allocated the same crop variety, or the variability in crop yields *between* the three different crop varieties themselves.

A typical set of data that might result from such an experiment is given in Table 7.1. The first step towards analysing these data is to create a basic plot (Figure 7.2). Figure 7.2a shows a lot of scatter: there is a large amount of variability in the dataset as

Table 7.1 Crop yield data for single factor analysis of variance.

Raw data treatment			Total sum of squares (SS_Y) $(y_i - \bar{y})^2$			Error sum of squares (SS_E) $(y_i - \bar{y}_i)^2$		
Crop A	Crop B	Crop C	A	B	C	A	B	C
9	16	8	4.41	24.01	9.61	0.00	3.24	4.41
11	14	8	0.01	8.41	9.61	4.00	0.04	4.41
10	15	12	1.21	15.21	0.81	1.00	0.64	3.61
9	13	13	4.41	3.61	3.61	0.00	1.44	8.41
6	12	11	26.01	0.81	0.01	9.00	4.84	0.81
10	15	10	1.21	15.21	1.21	1.00	0.64	0.01
10	16	9	1.21	24.01	4.41	1.00	3.24	1.21
5	17	11	37.21	34.81	0.01	16.00	7.84	0.81
9	9	8	4.41	4.41	9.61	0.00	27.04	4.41
11	15	11	0.01	15.21	0.01	4.00	0.64	0.81
n	10	10	10				36.0	49.6	28.9
Sum	90	142	101	SS_Y	264.7		SS_E	114.5	
Mean	9.0	14.2	10.1						

Overall n	30
Overall sum	333
Overall mean	11.1

$SS_Y - SS_E = SS_T$ 150.2

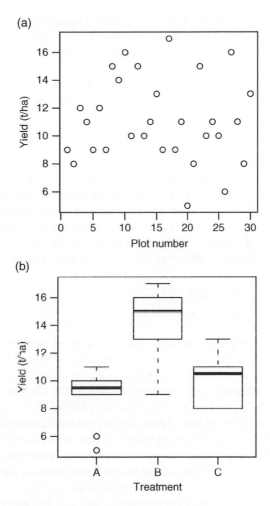

Figure 7.2 Plots of crop yields: (a) graph of yields by consecutive plot number; and (b) boxplot grouped by level.

a whole. When the data are organized by level (i.e. by crop variety) a pattern emerges: Crop B produces the best results on average but there are still some plots where Crops A and C did better than those where Crop B was planted. Our aim is to compare the three groups to see if the differences are statistically significant. That is, whether the differences are greater than would be expected by chance. Had there been only two levels, a t-test would have been sufficient, but in the case of three or more levels, it is necessary to compare the variances within and between treatments.

The overall amount of variability in the dataset is found by adding up the total sum of squares, SS_Y, which is given as:

$$SS_Y = \Sigma(y_i - \bar{y})^2, \qquad (7.1)$$

where y_i are the observations and $\bar{y}$ is the overall mean.

The quantity SS_Y is a good measure of the total variability in the dataset: it gets larger the further the observations are from the mean. Notice also that the total sum of squares is precisely the same as the numerator that was used in the formula for the variance in Chapter 2. In fact, dividing SS_Y by the number of degrees of freedom (df) gives the variance, just as before:

$$s_Y^2 = \frac{SS_Y}{df} = \frac{\Sigma(y - \bar{y})^2}{n-1}, \qquad (7.2)$$

In the example given here, the total sum of squares, SS_Y, is 264.7, as shown in Table 7.1. This variability is caused by many different factors: differences in nutrient availability, soil properties, microclimate, and – of most interest in the current investigation – the type of crop variety that was planted. Because we are interested in just the last of these factors we need to partition the variability between the crop yields into that which is attributable to the crop variety and that which is down to everything else. If the crop variety is an important factor, it will account for a large fraction of the total variability in the dataset, and the amount of variation within the plots that were planted with the same variety will be correspondingly small.

The amount of variability that remains after taking into account the choice of crop variety is called the residual variability. To calculate the residual variability, it is necessary to look at the squared differences between the yields in each plot and the means for each of the yields separately. That is, for Crop A find the sum of the squares of the differences between the yields of all plots which were assigned that crop. Such a procedure is shown in Table 7.1. If we then add up these differences for each of the treatment levels we arrive at a measure of the residual variability, which is called the error sum of squares, SS_E:

$$SS_E = \sum_{j=1}^{k} \sum \left(y_i - \overline{y}_j \right)^2, \tag{7.3}$$

in which we calculate the means for the jth of k treatments (in the present case there are three treatments, so $k = 3$) and then find the sum of squared differences within the treatments and add them all up to arrive at the SS_E value. In the present example, using the treatment means has reduced the variability (Figure 7.2b) and the value of SS_E calculated in Table 7.1 is 114.5.

Let us recap: the total sum of squares, SS_Y, measures the total amount of variability in the dataset, and the error sum of squares, SS_E, tells us how much variability is left over once the effect of the treatment is accounted for. We can now see that if the treatment makes no difference then SS_Y and SS_E will be the same. The bigger the difference between SS_Y and SS_E the more influence the treatment has on the differences between the measured values. We can calculate this difference, which we can call the treatment sum of squares, SS_T:

$$SS_T = SS_Y - SS_E \tag{7.4}$$

which in the current example is 264.7 − 114.5 = 150.2. Note that if the means were the same for the three groups then there would be no difference between SS_Y and SS_E − in other words, SS_T would be zero.

Bringing all of this information together gives Table 7.2, which is laid out in the typical format for an ANOVA table. Working out the number of degrees of freedom associated with each of the measures that we have calculated is very important, but it requires a bit of careful thinking. First, remember the rule from Chapter 2 that the number of degrees of freedom is the number of independent pieces of information that have been used to calculate the statistic. It can be worked out by taking the number of observations and subtracting the number of parameters that have been calculated. Let us take SS_Y, SS_E, and SS_T in turn. To calculate SS_Y we used only one statistic, the overall mean, and so we have to subtract one from the number of observations to find that the number of degrees of freedom associated

Table 7.2 ANOVA table for crop yield analysis of variance.

Source of variance	df	SS	MS	F	p
Treatment	2	150.2	75.1	17.7	0.000012
Error	27	114.5	4.2		
Total	29	264.7			

df, Degrees of freedom; SS, sum of squares; MS, mean square; F and p as defined in the text.

with SS_Y is $n - 1$ (Table 7.1). For SS_E, we calculated a mean for each of the k treatments. Each treatment mean has $n - 1$ degrees of freedom and so the total number of degrees of freedom associated with SS_E is $n - k = 27$ because for each of the k treatments we calculated a treatment mean and so we must reduce the number of degrees of freedom to account for the fact that we have used k treatment means to compute SS_E. For SS_T, k treatment means were calculated, but there are only $k - 1 = 2$ degrees of freedom because if we knew all but one of the treatment means we could calculate the remaining treatment mean using the overall mean.

Now that we have calculated the sum of squares and the degrees of freedom, the next step in performing the ANOVA test is to compute the mean squared values (abbreviated as MS). The mean squares, which are listed in Table 7.2 for the current example, are the equivalents of the variance (remember Equation 7.2), and they are calculated by dividing the sum of squares by the degrees of freedom for each row in the table.

The final step in calculating the result of the ANOVA test is to work out the F-statistic. Recall that our aim in conducting this test is to compare the treatment variance with the error variance. To make this comparison, we divide one by the other, to produce an F-ratio:

$$F = \frac{MS_{\text{Treatment}}}{MS_{\text{Error}}}, \tag{7.5}$$

In the current example, the F-ratio is 17.7, indicating that the variance between the treatment means is over 17 times greater than the variability that remains after the treatment means are accounted for. This is a good indication that the treatment means are different, but to be sure, we should work out the *p*-value for the test (i.e. calculate the probability of obtaining variances which differ by a factor of 17.7 or more by chance). This is a one-tailed test because we are only interested in knowing whether the treatment variance is greater than the error variance. The probability itself is obtained using the *F*-distribution (recall Section 5.4.5) with the number of degrees of freedom taken from the two variances that are being compared, which in this case are the treatment and error degrees of freedom (2 and 27, respectively). As with the other probability distributions, R can be used to calculate the probability as described below.

Worked Example 7.1 Plot the experimental crop yield data given in Table 7.1 using R, and analyse the data using the built-in ANOVA function.

Solution

Step 1: Read the data

It is recommended that ANOVA and other more advanced statistical procedures be performed using a specialist statistics package such as R. In R the process of reading in the data is as before, but it is important to make sure that you stick to the guidelines for producing dataframes because doing so will greatly simplify the subsequent analysis of data. In practice, this means making sure that the data are organized in two columns and *n* rows so that there is one row per sample. On each row there is a letter stating the treatment (A, B, or C), and a test score next to it. An example of such a datafile can be read in using the command:

```
cropyield <- read.csv("wex7_1.csv")
```

Step 2: Plot the data

Whilst a simple plot like Figure 7.2a can be created with the plot() command:

```
plot(cropyield$Yield, xlab="Plot number",
  ylab="Yield (t/ha)")
```

the boxplot() command produces more informative output especially when R's ability to organize the data by level is employed. In order that R can interpret the factor levels correctly, the column labelled 'Treatment' should be converted into a factor using:

```
cropyield$Treatment=as.
  factor(cropyield$Treatment)
```

A boxplot similar to that shown in Figure 7.2b can then be created using:

```
boxplot(Yield ~ Treatment, data=cropyield,
  xlab="Treatment", ylab="Yield (t/ha)")
```

in which the tilde symbol (~) indicates that we wish to split the variable named 'Yield' according to the groups given by the factor named 'Treatment'. The boxplot (Figure 7.2b) is much more informative and suggests that the mean of Crop B is larger than the other two, but that the ranges of variability are nevertheless quite broad.

Step 3: Calculate group means

The group means can be calculated easily using the `tapply()` function which gives R the powerful capability to apply any compatible function to subsets of a dataframe according to a factor variable. In the present case to calculate means, standard deviations, and counts of 'Yield' grouped by the factor 'Treatment' type:

```
tapply(cropyield$Yield, cropyield$Treatment,
  mean)
    A    B     C
 9.0  14.2  10.1
tapply(cropyield$Yield, cropyield$
  Treatment, sd)
       A         B          C
2.000000  2.347576  1.791957
```

```
tapply(cropyield$Yield, cropyield$Treatment,
length)
 A   B   C
10  10  10
```

Step 4: Perform ANOVA test

In order to perform the ANOVA calculations in R the command `aov()` is used, together with the `summary()` function which formats the output into the standard ANOVA table. The `aov()`

function uses R's model formula notation, which uses the tilde symbol (~). As with the boxplot example above, if we write 'Yield ~ Treatment', it is a shorthand way of saying: 'explain the variations in the variable "Yield" by using the variations in the variable "Treatment"'. To analyse the crop yield data, the following command would be used:

```
summary(aov(Yield ~ Treatment, data =
   cropyield))
```

which produces the standard ANOVA table with values identical to those calculate above and reported in Table 7.2.

```
            Df  Sum Sq  Mean Sq   F value      Pr(>F)
Treatment    2   150.2   75.100    17.709   1.221e-05 ***
Residuals   27   114.5    4.241
---
Signif. codes:  0 '***' 0.001 '**' 0.01 '*' 0.05 '.'
   0.1 ' ' 1
```

Note that R has calculated the p-value. To calculate this by hand based on the F-value of 17.709 with 2 and 27 degrees of freedom, the command is:

```
1 pf(17.709,2,27)

[1] 1.221034e-05
```

which gives the same answer. Note that in the ANOVA table, R indicates whether the test has passed a range of key significance levels. The analysis presented here indicates that even at the 0.001 significance level the null hypothesis should be rejected and we should conclude that the choice of crop variety has an effect on yield.

7.2 Assumptions and diagnostics

Before accepting and interpreting the results of any ANOVA, it is important to ensure that the assumptions that underlie the procedure have been met. Three conditions of particular note are:

1) The observations are independent of each other. This assumption can be checked by plotting the observations in the order that they were made in order to detect whether observations are systematically similar to (or consistently

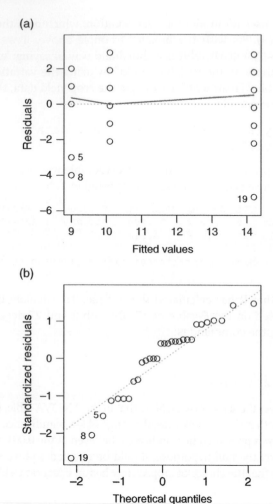

Figure 7.3 Diagnostic plots for analysis of variance: (a) residuals versus fitted values; and (b) normal probability plot of residuals.

different from) neighbouring values (Figure 7.3a). One of the most common reasons that this assumption might be violated is when repeated measures are made of the same individual. If this is an intentional part of the research design then specific repeated-measures ANOVA techniques are available (Montgomery, 2008).

2) The data within each treatment group are approximately normally distributed. The second assumption can be verified by using a normal probability plot of the residuals (Figure 7.3b), which may prompt the removal of outliers or transformation of the data. More detail on procedures for the latter is given in Chapter 8.

The relevant commands to produce these plots in R is:

```
plot(aov(Yield ~ Treatment, data =
    cropyield))
```

which produces a range of useful diagnostic plots, the first two of which correspond to those shown in Figure 7.3. In this case there are no notable departures from normality nor any dependencies among the residuals and so the analysis can proceed.

3) The samples have equal variances. This assumption is best checked using Barlett's test (which similar to the the *F*-test presented in Section 5.4.5, but for several samples). This procedure can be handled easily in R using the same syntax for model specification as was used in the aov() function:

```
bartlett.test(Yield ~ Treatment, data =
    cropyield)
```

In the present case the results are:

```
Bartlett test of homogeneity of vari-
    ances
data: Yield by Treatment
Bartlett's K-squared = 0.6393, df = 2, p-
    value = 0.7264
```

which, with a *p*-value of 0.73, confirms that variances are not significantly different.

7.3 Multiple comparison tests after analysis of variance

So far we have used ANOVA to tell us whether the variation among the treatment means is greater than the variation within the treatment groups themselves. The ANOVA procedure can only tell us whether there is a statistically significant difference between the treatment means. An obvious follow-up question is to ask which of the treatment means are different and by how much? For the reasons outlined at the beginning of this chapter, it is not possible to apply a series of successive *t*-tests to each of the treatments paired with the others in turn. Doing so would compound the Type I error probabilities of each *t*-test and therefore increase the overall chance of concluding that there was an effect when in fact there was not.

The solution to this problem is to use a multiple comparisons test. These are *post-hoc* tests and are designed to be carried out after ANOVA has confirmed an effect. There are two approaches in common use. The first is to use multiple *t*-tests but to adjust the critical *t*-value to maintain a total significance level of α. Each of the multiple *t*-tests is adjusted to use a significance level of α/n, so that if n tests are performed each with a significance level of α/n then the overall significance level is maintained at α. This procedure, which is known as the Bonferroni correction, is simple to understand and is guaranteed to maintain the overall significance level but it is very conservative and quickly becomes unworkable if the number of comparisons is high. Several alternatives have been proposed. For simple analyses, the Tukey test is the most useful practical alternative. Tukey's test is similar to the pairwise *t*-test but the test statistic is assumed to follow a distribution which is modified to account for multiple comparisons. Tukey's test (along with many other multiple comparison tests) can be performed readily in R using the function TukeyHSD(), as outlined in the following worked example which compares the different approaches to multiple comparisons.

Worked Example 7.2   Examine the data given in Table 7.1 which were analysed using the ANOVA procedure in Worked Example 7.1 to investigate which treatment means are different. Compare the results obtained using: (i) the unadjusted pairwise *t*-test; (ii) the Bonferroni corrected pairwise *t*-test; and (iii) Tukey's test.

Solution   A basic pairwise comparison of the means from this dataset can be obtained in R by using the function:

```
pairwise.t.test(cropyield$Yield,
  cropyield$Treatment, p.adjust="none")
```

Note that in this example, we have told R not to adjust the *p*-values for any of the tests (which is not a good idea in practice but which gives a good illustration of the differences). The results are:

```
Pairwise comparisons using t tests with
  pooled SD

data:   cropyield$Yield and cropyield$Treatment
   A         B
B 5.4e-06 -
C 0.24270 0.00013

P value adjustment method: none
```

The results are returned in table format where the row and column labels indicate which treatments are being compared. The results show that the *p*-value for a *t*-test on the means for Crop A and Crop C is high (i.e. the probability of getting a difference of this size by chance is high and so we should not reject the null hypothesis that the data are taken from the same sample). On the other hand, the results from the comparisons between Crops A and B, and B and C do indicate that the means for these crops are significantly different from each other at the 0.05 significance level, because the *p*-values are lower than 0.05.

Using the second method, namely the Bonferroni adjustment for *p*-values, requires the same R command but with a different setting for the 'p.adjust' option:

```
pairwise.t.test(cropyield$Yield, cropyield$
  Treatment, p.adjust="bonferroni")
```

It gives the following results:

```
Pairwise comparisons using t tests with
  pooled SD
data:   cropyield$Yield and
  cropyield$Treatment
   A         B
B 1.6e-05 -
C 0.7281  0.0004

P value adjustment method: bonferroni
```

Note that the same conclusion is reached, but the p-values (the lowest significance level at which we would reject the null hypothesis) are consistently higher, which means that there is a consistently higher probability that the differences in the means have occurred by chance rather than because the population means are different. Only the p-value for the comparison between Crop A and Crop C is greater than 0.05 and so in practice the conclusions of the study would not be altered.

To perform the Tukey test, the function in R is:

```
TukeyHSD(aov(Yield ~ Treatment, data =
   cropyield))
```

which gives the following output:

```
Tukey multiple comparisons of means

    95% family-wise confidence level

Fit: aov(formula = Yield ~ Treatment, data =
   cropyield)

$Treatment
       diff        lwr        upr       p adj
B-A     5.2   2.916581   7.483419   0.0000157
C-A     1.1  -1.183419   3.383419   0.4666129
C-B    -4.1  -6.383419  -1.816581   0.0003791
```

This function is slightly more informative in that it supplies the difference between the treatment means as well as an associated confidence interval (indicated by the abbreviations 'lwr' and 'upr') and the relevant p-value for each of the paired tests. The Tukey test gives p-values that lie between those from the uncorrected pairwise tests and the Bonferroni correction and offers a reliable approach for use in simple analyses.

The results of the multiple comparison tests show that at the 0.05 significance level, Crop B has a higher mean than either of Crops A and C but that the means for the latter two crop types are indistinguishable at this significance level. More extensive multiple comparisons are possible using orthogonal contrasts. These are beyond the scope of the present introductory text but should be considered in complex cases (Montgomery, 2008).

7.4 Non-parametric methods in the analysis of variance

The non-parametric equivalent of the one-way ANOVA is called the Kruskal–Wallis H-test. Like the other non parametric tests described in this book (e.g. the Mann–Whitney U-test), the Kruskal–Wallis H-test involves working with the ranked dataset. In fact, this test is almost a direct extension of the Mann–Whitney U-test but for three samples instead of two. The algebra is tedious but the procedure is available through the R function kruskal.test() in R and is invoked in precisely the same way as the standard ANOVA command aov().

Worked Example 7.3 A geographer is interested in finding out whether there is a difference in average pupil performance between schools in three different areas. In each of Areas A, B and C, eight schools are selected at random and information on the average A-level point scores is calculated (Table 7.3). Analyse the data using the Kruskal–Wallis H-test to determine whether average pupil performance differs between the three areas.

Table 7.3 Average A-Level point scores for schools in three areas.

Area A	Area B	Area C
673	666	850
671	834	917
903	930	858
589	785	723
1139	793	456
567	1048	749
699	707	723
858	778	610

Solution With a null hypothesis that there is no difference between the means of the three samples, and a significance level of 0.05, the Kruskal–Wallis *H*-test, which can be performed in R using:

```
schools <- read.csv("wex7_3.csv")

schools$Area=as.factor(schools$Area)

kruskal.test(Points ~ Area, data=schools)
```

and which produces the following results:

```
        Kruskal-Wallis rank sum test
data: Points by Area
Kruskal-Wallis chi-squared = 1.1297, df = 2,
  p-value = 0.5684
```

indicates via a *p*-value (the lowest significance level at which we would reject the null hypothesis) of 0.57 that there is no evidence to support the rejection of the null hypothesis and it should be concluded that there is no difference between the means of the three samples. The variability of average school performance within each area is greater than the variability between the areas, although a larger sample size would be of benefit were such a study performed in a research context.

7.5   Summary and further applications

In summary, the techniques introduced in this chapter allow the comparison of multiple samples, and open the way for the design and analysis of simple experiments. The general procedure for using analysis of variance can be broken down into five steps: (i) set out the experimental design including the number of treatments, samples within each treatment; (ii) allocate treatments to plots; (iii) compute sums of squares and degrees of freedom; (iv) calculate *F*-score and associated *p*-value; and (v) perform multiple comparisons tests as necessary.

The use of ANOVA to work with experimental data from the field and the lab is common in geographical studies. Several special cases that have not been covered in this section are worth noting because they arise frequently in geographical studies that use ANOVA. These situations include the presence of nuisance factors, and the study of experimental designs in which there is more than one factor that can be varied. These situations are described briefly here to warn the reader of their existence; the detailed mathematics of how to deal with these problems is beyond the scope of this book, but the reader is referred to more advanced texts on experimental design for details (e.g. Montgomery, 2008).

The first case concerns the situation that there is a nuisance factor also present in the sampling frame. In fact, the reason for randomization in the first place is to eliminate the possibility that nuisance factors can have an effect. If the nuisance factor itself can be measured, then it is possible to divide the sampling frame into 'blocks' in order to remove the effect of the nuisance factor. A clear example of the utility of blocking would arise were the sampling sites in the crop yield example of Section 7.1 split across two distinct soil types. In that case, yields might be affected by soil type at the same time as crop variety. By splitting the available sites into groups of the same soil type and randomizing the experiment within these blocks, the problem of the nuisance factor (i.e. soil type) can be eliminated. In the design of experiments, the general advice is that it is wise to block what you can and randomize what you cannot.

A natural extension of the problem of blocking is two-way ANOVA. This technique is used in situations where the analyst is interested in the joint effects of two factors on the results of an experiment rather than just one. An example would arise if in addition to crop variety, the investigator wished to discover whether amount of irrigation also played a role in governing crop yield. In this case, the crop variety and irrigation regime could be randomly assigned and variability arising from each factor partitioned in an ANOVA table.

Exercises

1 Repeat the analysis from Worked Example 7.1 using the Kruskal–Wallis *H*-test and compare the results.

2 A geomorphologist wishes to determine whether rock type controls slope angle and takes a random sample of measured slope angles from schist, slate and mudstone slopes, given in

Table 7.4. Analyse these data to determine whether slope angle differs on slopes of different lithology.

Table 7.4 Slope angles and lithology.

Schist	Slate	Mudstone
43	34	16
44	28	23
44	29	24
36	35	18
38	34	24
27	28	17
36	28	24
43	31	23
37	37	24
37	32	23

3 A geographer is interested in literacy levels in a city and wishes to determine whether there is any variation in the levels of attainment amongst school children across the city. Suggest an appropriate analysis and sampling design and state any assumptions that you are required to make.

8

Correlation

STUDY OBJECTIVES

- Calculate and interpret Pearson's product-moment correlation coefficient, r_{xy}, and understand the assumptions that lie behind its use.
- Know how to test for the significance of r_{xy} and interpret the result.
- Understand the reasons why a non-parametric correlation measure is sometimes appropriate and in those cases be able to calculate Spearman's rank correlation coefficient and interpret the result.
- Recognize the limitations of correlation analysis including the principle that correlation does not imply causality, and understand other common pitfalls that may arise in correlation studies.

8.1   Correlation analysis

Up to this point the analyses presented in this book have concentrated on the detection of differences between populations. In this chapter, the question of the relation between two datasets is addressed. The association between two variables can be measured using the correlation coefficient. A positive correlation indicates that as one variable increases so the other variable increases. A negative correlation indicates the opposite: that as the value of one variable increases the value of the other variable decreases. When the correlation coefficient is

Statistical Analysis of Geographical Data: An Introduction, First Edition.
Simon J. Dadson.
© 2017 John Wiley & Sons Ltd. Published 2017 by John Wiley & Sons Ltd.

zero, there is no association between the variables. The value of the correlation coefficient lies between −1 and +1, with positive values indicating a positive correlation and negative values a negative correlation.

8.2 Pearson's product-moment correlation coefficient

The most common measure of correlation is the Pearson product-moment correlation coefficient. It is based on the covariance, which is a more general form of the variance (see Section 2.3.7). Whereas the variance measures the variability *within* a single dataset, the covariance measures the variability *between* two datasets. The covariance is defined as:

$$\text{cov} = \frac{\sum_{i=1}^{n}(x_i - \bar{x})(y_i - \bar{y})}{n-1}, \tag{8.1}$$

where x and y are the two variables in question. The covariance of x with itself is identical to the variance. Both the variance and the covariance have units which are the square of the units of the original variable and so to obtain a meaningful measure that permits comparisons between datasets and has a finite range, we divide the covariance by the standard deviation of each of the individual datasets to give the correlation coefficient:

$$r_{xy} = \frac{\sum_{i=1}^{n}(x_i - \bar{x})(y_i - \bar{y})}{(n-1)s_x s_y}, \tag{8.2}$$

which can be rearranged to provide a more convenient form for calculation by hand:

$$r_{xy} = \frac{\left(\sum_{i=1}^{n} x_i y_i\right) - n\bar{x}\bar{y}}{(n-1)s_x s_y}. \tag{8.3}$$

Worked Example 8.1 This example is based on a classic problem given by Draper and Smith (1981) in their book *Applied Regression Analysis*. It has been updated (and converted to metric units) using data from the United States Department of Agriculture National Agricultural Statistics Service (http://www.nass.usda.gov), which were also plotted graphically in Figure 2.4. Table 8.1 gives July rainfall totals and yields in the US state of Iowa between 1930 and 1962. Calculate Pearson's product-moment correlation coefficient, r_{xy}. What does the result tell you about the relation between rainfall and wheat yield?

Table 8.1 July rainfall and corn yield in Iowa, USA.

Year	July rainfall (mm)	Yield (t/ha)
1930	38	2.1
1931	69	2.1
1932	79	2.7
1933	88	2.5
1934	98	1.4
1935	85	2.4
1936	13	1.3
1937	67	2.8
1938	108	2.9
1939	80	3.3
1940	116	3.3
1941	57	3.2
1942	124	3.8
1943	116	3.4
1944	95	3.3
1945	75	2.7
1946	62	3.6
1947	44	1.9
1948	105	3.8
1949	88	2.9
1950	118	3.0
1951	113	2.7
1952	98	3.9

(Continued)

Table 8.1 (Continued)

Year	July rainfall (mm)	Yield (t/ha)
1953	83	3.3
1954	45	3.4
1955	84	3.0
1956	115	3.3
1957	90	3.9
1958	192	4.1
1959	58	4.0
1960	70	4.0
1961	140	4.7
1962	159	4.8

Source: Data from United States Department of Agriculture National Agricultural Statistics Service, http://www.nass.usda.gov.

Solution

This question asks for the correlation coefficient. If this is computed by hand, then using the computational formula given in Equation 8.3 can save a substantial amount of time. This formula requires the calculation of the sum of the product of each pair of observations, as shown in Table 8.2.

The sum of the products of the paired observations, $\sum_{i=1}^{n} x_i y_i$, is 9859.8 which when put into Equation 8.3 gives a value for r_{xy} of 0.57. The correlation is positive, which indicates that when July rainfall is high, corn yields also tend to be high.

Excel gives the same answer using its function =correl(), and the same value can be obtained using R with:

```
iowacorn <- read.csv("wex8_1.csv")
cor(iowacorn$Rain,iowacorn$Yield)
```

8.3 Significance tests of correlation coefficient

The question of whether the correlation coefficient is significantly different from zero can be addressed using the hypothesis testing framework introduced in Section 5.1.3.

Table 8.2 Calculations for Pearson's product-moment correlation coefficient between July rainfall and corn yield in Iowa, USA.

Year	July rainfall (mm)	Yield (t/ha)	Rainfall × Yield
1930	38	2.1	79.8
1931	69	2.1	144.9
1932	79	2.7	213.3
1933	88	2.5	220.0
1934	98	1.4	137.2
1935	85	2.4	204.0
1936	13	1.3	16.9
1937	67	2.8	187.6
1938	108	2.9	313.2
1939	80	3.3	264.0
1940	116	3.3	382.8
1941	57	3.2	182.4
1942	124	3.8	471.2
1943	116	3.4	394.4
1944	95	3.3	313.5
1945	75	2.7	202.5
1946	62	3.6	223.2
1947	44	1.9	83.6
1948	105	3.8	399.0
1949	88	2.9	255.2
1950	118	3.0	354.0
1951	113	2.7	305.1
1952	98	3.9	382.2
1953	83	3.3	273.9
1954	45	3.4	153.0
1955	84	3.0	252.0
1956	115	3.3	379.5
1957	90	3.9	351.0
1958	192	4.1	787.2
1959	58	4.0	232.0

(Continued)

Table 8.2 (Continued)

Year	July rainfall (mm)	Yield (t/ha)	Rainfall × Yield
1960	70	4.0	280.0
1961	140	4.7	658.0
1962	159	4.8	763.2
Sum	2972	103.5	9859.8
Mean	90.06	3.136	
Std dev.	35.54	0.8295	

An appropriate null hypothesis is H_0: $r_{xy} = 0$, which can be tested using the statistic:

$$t_{calc} = \frac{r\sqrt{n-2}}{\sqrt{1-r^2}}, \tag{8.4}$$

which follows a t-distribution with $n-2$ degrees of freedom. For a given level of significance, α, the null hypothesis can be rejected only if $t_0 > t_{\alpha/2,n-2}$.

Worked Example 8.2 Test the hypothesis that the correlation coefficient calculated in Worked Example 8.1 is significantly different from zero. Use a significance level of 0.05.

Solution
With the null hypothesis and significance level as above, the value of t_{calc} is 3.87, whilst t_{crit} is 2.04 (with $\alpha = 0.05/2$, and degrees of freedom $= 33 - 2 = 31$), so it is possible to reject the null hypothesis and conclude that the correlation coefficient is different from zero at the 0.05 significance level.

8.4 Spearman's rank correlation coefficient

When data are not normally distributed Spearman's rank correlation coefficient may be used. As with many non-parametric statistics this measure is calculated using ranks rather than the data values themselves. Spearman's rank correlation is more robust than Pearson's correlation coefficient in the presence of outliers. Moreover, it is of wider applicability than Pearson's method because both ordinal scale and interval scale data can be used. The first step in the procedure is to rank interval-scale data within each variable so that they are on an ordinal scale. If the data are already in ranked form, then this first step is not required. Where ranks are tied, the mean of the ranks that the tied observations would have received is used.

The next step is to find the difference between the ranks of the corresponding observations in each dataset. If the two datasets are positively correlated, then the ranks will tend to be more alike than if they are negatively correlated. If there is no correlation, then there will be no apparent pattern to the differences in the ranks. The Spearman's rank correlation coefficient, r_S, is therefore defined as:

$$r_s = 1 - \frac{6 \Sigma d^2}{n\left(n^2 - 1\right)}, \tag{8.5}$$

where d is the difference between each pair of ranks. Spearman's correlation coefficient, r_S takes values between -1 and 1, with negative values indicating negative correlations. The procedure for testing the significance of this correlation coefficient is identical to the procedure developed in Section 8.3 for Pearson's r (Zar, 1972).

Worked Example 8.3 Use Spearman's rank correlation coefficient to evaluate the strength of the association between gross domestic product (GDP) per capita and life expectancy using the global dataset plotted in Figure 2.5.

Solution
The calculations can be done manually but are too cumbersome to report here given the size of this dataset. The dataset is available in the file gdp_lifeexp.csv:

```
gdplife <- read.csv("wex8_3.csv")
```

The cor() function in R can be modified to calculate Spearman's correlation coefficient as follows:

```
cor(gdplife$GDP_percap, gdplife$Life_exp,
   method="spearman")
```

which gives the result $r_S = 0.84$, that is, a strong positive correlation: as a country's GDP increases so does the life expectancy of its inhabitants. The value of t_{calc} associated with this correlation coefficient is 20.8 which against a t_{crit} of 1.97 (with $\alpha = 0.05/2$, and degrees of freedom $= 178 - 2 = 176$) indicates a significant correlation.

It is worth noting that Spearman's correlation is a sensible choice for use with this dataset because the GDP data in particular are highly skewed and do not follow anything like a normal distribution. Had Pearson's method been used the r_{xy} value would be substantially lower than the r_S value obtained in the present example (although it would still have been statistically significant). This is because Pearson's method accounts only for *linear* association between the two variables. If the association is non-linear, as it is in the present example, Spearman's method can prove superior providing that the true relation between the variables is monotonic.

8.5 Correlation and causality

Correlation analysis is a particularly powerful technique because it can be used to explore the links between processes and their effects, in both human and physical geography. Yet the interpretation of the results of correlation analysis requires caution. Correlation should be taken as nothing more than a statement about shared variability in two datasets and whilst a statistically significant positive or negative correlation may prompt further investigation it is important to be alert to a range of potential misinterpretations that can arise.

First, correlation does not imply causality; nor is it possible to infer the direction of potential causality from a statement of correlation. A positive correlation between the number of ice creams sold at the seaside and the number of hours of sunshine should not lead to the conclusion that selling ice creams causes the sun to come out.

Second, the presence of one or several 'hidden variables' might create misleading correlations. For example, a strong positive correlation between the number of fire-fighters sent to tackle a blaze and the number of people that die in the fire does not mean that fire-fighters are causing deaths. Instead there is a hidden variable, in this case the size of fire, with which both of the other variables are correlated. The correlation is a perfectly valid statement of the association between the two variables, and might even be statistically significant, but there is no causality associated with it. A more serious, scientific example was reported by Quinn *et al.* (1999) from an investigation into the development of myopia (short-sightedness) in 479 children. Children who slept with the light on before the age of two were found to have a much greater incidence of myopia later in life. Subsequent studies argued that parents who were themselves myopic were much more likely to use a night light and that parental myopia was a much more likely reason for the development of myopia in children (Gwiazda *et al.*, 2000; Zadnik *et al.*, 2000). Here parental myopia is the 'hidden variable' with which both night light use and myopia in offspring are correlated.

Third, spurious correlations can arise in situations where many different correlations are calculated across a range of variables. Because in any calculation of the correlation coefficient there is a finite probability that a correlation might be found by pure chance (i.e. the Type I error), if many correlations are calculated then the Type I error probabilities are compounded. The calculation of many correlations between variables is unfortunately made much easier by the availability of computer software but it should be discouraged without reference to a body of theory against which to evaluate the correlations that might be observed.

Fourth, small sample sizes can lead to misleadingly high values of the correlation coefficient, and large samples can contain important correlations that nevertheless do not yield impressive values of the correlation coefficient. Caution must be exercised in interpreting correlation coefficients from samples of different sizes, and it is important always to test for the significance of the correlation rather than relying on 'rules of thumb' about what is a good correlation and what is not.

Exercises

1   Repeat the correlation analysis presented in Worked Example 8.1 using Spearman's rank correlation coefficient and discuss the differences that arise.

2   Property prices are thought to vary with distance from a city centre. Use the data presented in Table 8.3, which are taken from a random sample of property values in a city, to test this theory. State the assumptions that you have made in your analysis.

Table 8.3 Property prices in Oxford, UK.

Distance (km)	Value (£)
1.4	263 000
1.9	238 000
2.2	202 000
2.6	214 000
3.5	189 000
4.2	178 000
4.6	176 000
4.8	189 000

Source: Aggregated from data produced by Land Registry © Crown copyright 2012.

3   Many studies of island biogeography have found that the number of species found on an island increases with the size of the island. Having selected an appropriate correlation measure, analyse the data in Table 8.4 in order to test this hypothesis. (Hint: plot the data first, to evaluate whether the assumptions behind your chosen correlation method are met.)

Table 8.4 Species–area relationship for birds of the Solomon Archipelago.

Island	Area (km^2)	Number of species
Bougainville	8591	82
Guadalcanal	5281	79
Ysabel	4095	71
Choiseul	2966	70
Buka	611.2	63
Nggela (Florida)	367.8	61
Shortland (Alu)	231.8	58
Fauro	71.0	51
Vatilau (Buena Vista)	14.0	41
Nusave	0.534	35
Bagora	0.327	22
Nugu	0.149	9
Samarai	0.0909	19
New	0.0707	15
Dalakalonga	0.0666	10
Elo	0.0546	14
Kosha	0.0515	13
Kukuvalu	0.0404	11
Tapanu	0.0161	13
Kanasata	0.0091	6
Nameless	0.0070	8

Source: Diamond and Mayr, 1976.

9

Linear regression

STUDY OBJECTIVES

- Understand the motivation for linear regression and the basic geometric and algebraic description of a linear model.
- Know the six key steps involved in regression analysis together with the reasons underlying each step.
- Be able to calculate the coefficients of a linear model and to create and interpret diagnostic plots of the resulting model and its residuals.
- Understand how to calculate the significance of the regression based on an analysis of variance (ANOVA) approach and to calculate and interpret the coefficient of determination.
- Compute confidence intervals on the regression parameters and determine whether the regression is useful.
- Know when regression assumptions are not valid and be able to use appropriate alternative techniques such as simple transformations to perform basic non-linear regressions, and reduced major axis regression when error is present in both variables.

9.1   Least-squares linear regression

In the previous chapter, techniques were introduced to explain the association between two variables. The next step is to quantify the relation between variables in order to make predictions of one variable given the value of another. In order to make

Statistical Analysis of Geographical Data: An Introduction, First Edition.
Simon J. Dadson.
© 2017 John Wiley & Sons Ltd. Published 2017 by John Wiley & Sons Ltd.

predictions it is necessary to create a model, that is to develop a mathematical relation which links the two variables, x and y. The most basic of these relations is the linear model:

$$y = mx + c, \tag{9.1}$$

which represents a straight line with slope, m, and intercept, c. Here x is the independent variable and y the dependent variable. Once values of m and c are known the model can be used to predict a value of y for any given value of x. Linear regression is the name given to the procedure used to determine appropriate values of m and c given a set of observed data. Linear regression is a very common technique and can be divided into six steps:

1) Produce initial exploratory plots of the data.
2) Decide on the appropriate formulation of the model.
3) Calculate the coefficients of the linear model.
4) Interpret diagnostic plots of residuals.
5) Calculate the significance of the regression and coefficient of determination.
6) Compute confidence intervals on the regression parameters.

9.2   Scatter plots

The most useful initial step in a regression analysis is to produce a scatter plot showing the association between the dependent and independent variables. Plotting the data prior to analysis also permits the identification of several problematic features which if present would produce misleading results. These include outliers, non-linear behaviour, heteroscedasticity, and departure from normality, which will be examined in detail in this chapter.

Worked Example 9.1   An ecologist wishes to estimate the height of trees using only measurements of their girth (which are easier to collect). The ecologist measures tree girth and height for 31 felled trees and wishes to fit a linear model to the data (if appropriate). Using the built-in R dataset named 'trees' (Table 9.1) make a plot of tree height against girth. Note that the built-in dataset is originally from Atkinson (1982) and contains data on a sample of 31 trees measured in the USA.

Table 9.1 Tree height and girth data.

Height (m)	Girth (m)
21.34	0.211
19.82	0.218
19.21	0.224
21.95	0.267
24.70	0.272
25.30	0.274
20.12	0.279
22.87	0.279
24.39	0.282
22.87	0.284
24.09	0.287
23.17	0.290
23.17	0.290
21.04	0.297
22.87	0.305
22.56	0.328
25.91	0.328
26.22	0.338
21.65	0.348
19.51	0.351
23.78	0.356
24.39	0.361
22.56	0.368
21.95	0.406
23.48	0.414
24.70	0.439
25.00	0.445
24.39	0.455
24.39	0.457
24.39	0.457
26.52	0.523

Source: Atkinson, 1982. Reproduced with
    permission of Wiley.

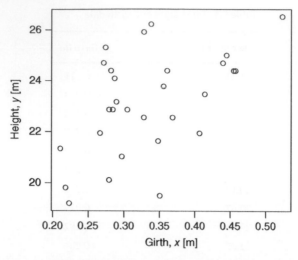

Figure 9.1 The relation between tree height and diameter in 31 cherry trees. Source: Atkinson, 1982. Reproduced with permission of Wiley.

Solution   A scatter plot can be created easily in Excel by using the chart wizard. In R the command is `plot ()`, a plot of R's built-in 'trees' dataset showing height versus diameter can be produced using:

```
library(datasets)

girth <- trees$Girth/39.37 # convert from inches to metres
height <- trees$Height/3.28 # convert from feet to metres

plot(girth, height, xlab="Girth, x [m]",
ylab="Height, y [m]")
```

noting that this particular dataset is given in feet and inches and has, in the above example, been converted to metres (Figure 9.1).

9.3   Choosing the line of best fit: the 'least-squares' procedure

The aim of linear regression is to select the straight line that provides the best fit to the data. The line of best fit can be chosen using any suitable method. The simplest method would be to sketch by hand a line of best fit and then to calculate its slope, m, using the formula slope = vertical difference/horizontal

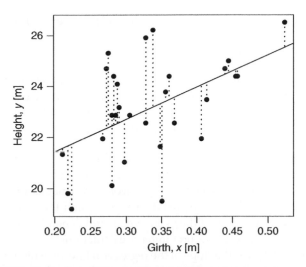

Figure 9.2 Illustration of the principle behind linear least squares regression.

difference (or rise/run) and its *y*-intercept, *c*, by examining graphically where the fitted line crosses the vertical axis. The problem with this approach is that the values obtained for the parameters would depend on precisely how the line of best fit was drawn, and that might vary from person to person. An objective way to select the line of best fit is required. The best fit line must pass through the 'middle' of the dataset, defined as the point $(\bar{x}, \bar{y})$. The angle of the line is chosen so that the squares of the distances between each of the points and the line are kept to a minimum (see the dotted lines in Figure 9.2). Note that the squares of the distances are used so that equal account can be taken of negative and positive departures from the line of best fit, although this process does mean that the method can be sensitive to extreme outliers. A regression line that has been fitted in this way is called a least squares regression line and the values of the slope and intercept parameters *m* and *c* are uniquely determined.

The slope of the line of best fit (*m* in Equation 9.1) can be calculated from the data using the following formula:

$$m = \frac{(\sum xy) - n\bar{x}\bar{y}}{\sum x^2 - n\bar{x}^2}, \qquad (9.2)$$

Once m has been calculated, the intercept, c, can be found:

$$c = \bar{y} - m\bar{x}. \tag{9.3}$$

Worked Example 9.2 Fit a linear regression model to predict tree height given tree girth from Table 9.1 using the least-squares procedure to estimate the parameters m and c. Add the regression line to the scatter plot from Worked Example 9.1.

Solution The calculations are straightforward, if a little cumbersome, to perform by hand. In addition to standard estimates of the mean of each dataset, the steps involve the calculation of the sum of squares for the independent variable and the sum of the products of the values of the independent and dependent variable (Table 9.2). Once these preliminary calculations are complete, Equation 9.2 gives the slope of the regression line, m, which in this case is 12.7. With that information, the intercept can be computed using Equation 9.3. In this case $c = 18.9$, so the final fitted model is:

$$\text{tree height} = 18.9 + 12.7 \times \text{tree girth} \tag{9.4}$$

The same calculations can be performed in Excel using the built-in functions `=slope()` and `=intercept()`, which give the same results. In R, the fitted values can be calculated using the linear model function, `lm()` which takes an argument in model notation (see Section 7.1), so `lm(y~x)` fits a linear model using y as the dependent variable and x as the independent variable:

```
treemodel <- lm(height ~ girth)
```

which gives the results that are identical to those obtained by hand, and saves the linear model as an object called 'treemodel' for subsequent use:

```
Call:

lm(formula = height ~ girth)

Coefficients:
(Intercept)          girth
      18.91          12.66
```

To add the regression line to the scatterplot (as in Figure 9.2) use the R function `abline()`. Note that it is possible to use extract the

Table 9.2 Calculations to estimate parameters of best-fit linear model to trees data using least squares procedure.

	Height	Girth			
	y	x		x×y	x^2
	21.34	0.211		4.5027	0.0445
	19.82	0.218		4.3208	0.0475
	19.21	0.224		4.3030	0.0502
	21.95	0.267		5.8607	0.0713
	24.70	0.272		6.7184	0.0740
	25.30	0.274		6.9322	0.0751
	20.12	0.279		5.6135	0.0778
	22.87	0.279		6.3807	0.0778
	24.39	0.282		6.8780	0.0795
	22.87	0.284		6.4951	0.0807
	24.09	0.287		6.9138	0.0824
	23.17	0.290		6.7193	0.0841
	23.17	0.290		6.7193	0.0841
	21.04	0.297		6.2489	0.0882
	22.87	0.305		6.9754	0.0930
	22.56	0.328		7.3997	0.1076
	25.91	0.328		8.4985	0.1076
	26.22	0.338		8.8624	0.1142
	21.65	0.348		7.5342	0.1211
	19.51	0.351		6.8480	0.1232
	23.78	0.356		8.4657	0.1267
	24.39	0.361		8.8048	0.1303
	22.56	0.368		8.3021	0.1354
	21.95	0.406		8.9117	0.1648
	23.48	0.414		9.7207	0.1714
	24.70	0.439		10.8433	0.1927
	25.00	0.445		11.1250	0.1980
	24.39	0.455		11.0975	0.2070
	24.39	0.457		11.1462	0.2088
	24.39	0.457		11.1462	0.2088
	26.52	0.523		13.8700	0.2735
Mean	23.17	0.337			
Count	31		Sum	244.158	3.702

values of m and c from the variable named 'coefficients' in the saved object 'treemodel':

```
abline(treemodel$coefficients[1],
   treemodel$coefficients[2])
```

9.4 Analysis of residuals

The adequacy of the fitted model can be evaluated in several ways, the first of which is to look at the residuals. The residuals are the differences between the observed and predicted values of the dependent variable, represented as the dotted lines in Figure 9.2. In fact, it was the squares of the residuals that were minimized during the least-squares fitting procedure. Several properties of the regression model can be evaluated by looking at the residuals and this is typically the first diagnostic that the analyst should use, because there is little point trying to use more sophisticated measures of goodness of fit if one or more of the assumptions underlying the regression model are not met.

First, if the residuals are large compared with the overall amount of variability in the original dataset then it is apparent that the regression model does not explain much of the variability in the observations. In Section 9.6, a quantitative measure of the amount of variability explained by the model will be developed. Second, under normal circumstances there should be no pattern to the residuals. If a pattern is detected it may indicate either a problem with the data or a problem with the form of the fitted model (e.g. if the data would be better matched with a curved line rather than a straight one). Third, a plot of the residuals should be used to confirm that the residuals are normally distributed, and that the variance of the residuals is uniform across the range of the dataset. A fourth benefit of investigating residuals is to determine whether specific data points can be classified as outliers. Outliers are typically defined as points that are more than 3 standard deviations above or below the regression line (on the basis that roughly 99% of data points should lie within this interval if the residuals are normally distributed). It is, however, important to investigate outliers in

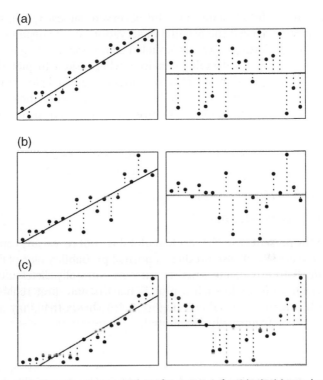

Figure 9.3 Plots showing typical configurations of residuals: (a) standard, normally distributed, homoscedastic; (b) heteroscedastic; and (c) non-linear model.

detail to determine whether they have arisen as a result of a simple problem with data entry (in which case they might be corrected) or if they are indicative of a peculiar example that can be explained using the specific circumstances of an individual case. Some typical plots of residuals are shown in Figure 9.3.

Worked Example 9.3   Calculate residuals for the regression developed in Worked Example 9.2 and plot them. Comment on any pattern present.

Solution   To calculate the residuals, it is first necessary to use the fitted model (i.e. Equation 9.4) to obtain the fitted value, $\hat{y}_i$, which corresponds to each of the observations x_i. The differences between the fitted values and the observed values, $y_i - \hat{y}_i$, are then plotted against the original observations of the dependent variable, y_i. Plotting residuals is such a commonplace diagnostic procedure that R's default behaviour when asked to plot the 'treemodel' regression object is to produce a residual plot, along with a series of more advanced diagnostic plots. Nevertheless, a customized residual plot (e.g. Figure 9.4a) can still be produced using:

```
plot(treemodel$fitted, treemodel$resid,
  xlab="Fitted value", ylab="Residual")
```

```
abline(0,0)
```

```
hist(treemodel$resid)
```

In the present case, the residual plot does not reveal any cause for concern. R will also produce a normal probability plot of the residuals to investigate whether they are normally distributed, although with $n < 40$ such a plot is hard to interpret reliably. The histogram of residuals (Figure 9.4b) shows that they are approximately normally distributed.

9.5   Assumptions and caveats with regression

The diagnostic residual plots shown in the previous section can provide useful indicators that the assumptions of regression have been met. A formal statement of these assumptions includes: (i) that the values of the independent variable are known without error – in practice this assumption is often not met, but if the error in measuring the independent variable is much less than the error in the dependent variable it is often sufficient to merit continuation with the regression procedure; and (ii) the residuals are independent of each other and are normally distributed. Failure to meet the last assumption can often be corrected through the use of variance-stabilizing transformations which are described in more advanced books on regression (e.g. Draper and Smith, 1981).

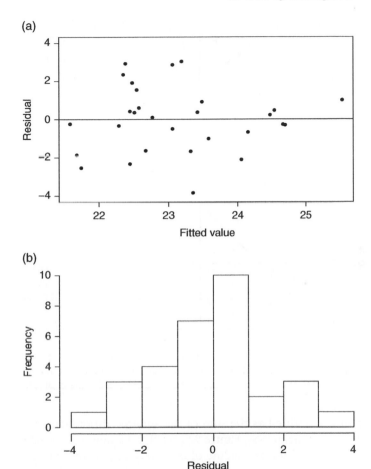

Figure 9.4 Diagnostic plots using residuals for tree height regression: (a) residuals against fitted values; and (b) histogram of residuals.

9.6 Is the regression significant?

In order to determine whether the line of best fit accounts for more of the variability in the original dataset than it leaves behind, it is necessary to calculate the proportion of the total variability explained by the line of best fit. Three quantities are required: (i) the amount of variability to begin with, which

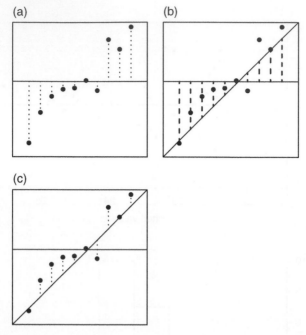

(a)

(b)

(c)

Figure 9.5 Regression by linear least squares: (a) the spread around the mean in the vertical direction; (b) the spread accounted for by the line of best fit; and (c) the residual deviation, or spread around the line of best fit.

can be quantified as the sum of the squared differences from the overall mean (SS_Y; Figure 9.5a); (ii) the amount of variability accounted for by the line of best fit, which is measured as the sum of squared differences between the fitted values and the mean (SS_{Reg}; Figure 9.5b); and (iii) the amount of variability left over after the model has been fitted, which is calculated as the sum of the squared differences between each of the data points and the corresponding point on the line of best fit (SS_E; Figure 9.5c). The total variance can be divided into a component which is due to the fitted regression line, and a component which represents the residual variation of data around the regression line. In detail, this equation can be rewritten as:

$$SS_Y = SS_{Reg} + SS_E \tag{9.5}$$

where the individual components are:

$$\sum_{i=1}^{n}(y_i - \bar{y})^2 = \sum_{i=1}^{n}\left(\hat{y}_i - \bar{y}\right)^2 + \sum_{i=1}^{n}\left(y_i - \hat{y}_i\right)^2. \tag{9.6}$$

It is possible to calculate these components individually, although if doing the calculations by hand it is useful to note that

$$SS_Y = \sum_{i=1}^{n}(y_i - \bar{y})^2 = \left(\sum_{i=1}^{n}y_i^2\right) - n\bar{y}^2, \tag{9.7}$$

and also that

$$SS_E = SS_Y - mS_{xy}, \tag{9.8}$$

where S_{xy} is the corrected sum of products (i.e. the numerator in Equation 9.2).

It should be noted that there is a very important similarity between these quantities and their equivalents that were used in the ANOVA in Chapter 7. In fact, it is common practice to set out the components of the variances associated with the linear model used in regression analysis in an ANOVA table as shown previously (e.g. Table 7.2). The sums of squares are converted into variances by dividing by the number of degrees of freedom: SS_Y has $n-1$ degrees of freedom, SS_{Reg} has one degree of freedom, and SS_E has $n-2$ degrees of freedom. The ratio of the regression variance to the residual variance follows an F-distribution:

$$F_{calc} = \frac{MS_{Reg}}{MS_E}, \tag{9.9}$$

and can be compared with a critical value (with one and $n-2$ degrees of freedom) from Table B.4 in Appendix B in order to determine whether the amount of variability explained by the linear model is greater than the remaining variability by a statistically significant amount. The null hypothesis that the slope of the regression line is zero can be rejected if $F_{calc} > F_{\alpha,1,n-2}$, where α is the significance level.

Worked Example 9.4 Calculate the components of variance attributable to regression and residuals in Worked Example 9.3, and use these results to perform an *F*-test in order to determine whether the regression is significant.

Solution The computational formulae given in Equation 9.7 and Equation 9.8 greatly simplify the manual computation of the components of variance involved in regression analysis. The key intermediate calculations are for the sum of squares of *y* and the sum of products of *x* and *y*. The latter has already been calculated in Table 9.2 and the former can be obtained in a similar way:

$$SS_Y = \sum_{i=1}^{n}(y_i - \bar{y})^2 = \left(\sum_{i=1}^{n} y_i^2\right) - n\bar{y}^2 = 16757.32 - (31)(23.17129)^2$$
$$= 113.151, \quad (9.10)$$

$$SS_E = SS_Y - mS_{xy} = 113.151 - (12.661)(2.412) = 82.613, \quad (9.11)$$

$$SS_{Reg} = SS_Y - SS_E = 113.151 - 82.612 = 30.538. \quad (9.12)$$

Note that in the intermediate steps of the calculations it is necessary to keep track of values to a far greater precision than will be reported in the final answer. This point is especially important when subtracting two numbers of similar magnitude. The resulting ANOVA table is shown in Table 9.3.

To test the significance of the regression the calculated *F*-score is compared with the critical value (with $\alpha = 0.05$ and 1 and 29 degrees of freedom), which is 3.35. Since the calculated *F*-score is greater than the critical value the null hypothesis that the amount of variability explained by the regression model is the same as the amount of variability left behind can be rejected

Table 9.3 ANOVA table for regression between tree height and diameter.

Source	df	SS	MS	F
Regression	1	30.54	30.54	10.71
Residual	29	82.61	2.85	
Total	30	113.15		

df, Degrees of freedom; SS, sum of squares; MS, mean square; *F*, *F*-statistic.

and it is concluded that the regression model explains a statistically significant fraction of the variability in the data.

The components of variance can easily be extracted in R using the `summary.aov()` function together with linear model that was fitted in Worked Example 9.2:

```
summary.aov(treemodel)
           Df Sum Sq Mean Sq F value    Pr(>F)
girth       1 30.528 30.5283  10.707  0.002758 **
Residuals  29 82.686  2.8512
---
Signif. codes:  0 '***' 0.001 '**' 0.01 '*' 0.05 '.'
0.1 ' ' 1
```

The ANOVA table and the conclusions are almost identical, the only difference being very minor due to round-off error in the manual calculation of the residual sum of squares.

9.7 Coefficient of determination

The fraction of the total amount of variability that is explained by the model is given by the coefficient of determination, R^2, which is defined as:

$$R^2 = \frac{SS_{Reg}}{SS_Y} = 1 - \frac{SS_E}{SS_Y}. \tag{9.13}$$

It is numerically equivalent to the square of the correlation coefficient (see Section 8.2). The R^2 value is very commonly quoted in statistical analyses, but it is often misinterpreted. It is important to realize that R^2 does not give any information on the *strength* of the relation between the two variables (i.e. whether a change in the independent variable causes a large or a small change in the dependent variable). Nor does the value of R^2 indicate whether the fitted model is appropriate: it may be that the relation between the two variables is in fact not linear at all, and a high R^2 gives no guarantee that the model will make useful predictions of future observations.

One particular difficulty with R^2 is that it is partly dependent on sample size. Caution must be exercised when comparing R^2 values from studies with different sample sizes. The most straightforward way to account for this problem is to test for the significance of R^2 using the methods discussed for testing the correlation coefficient in Section 8.3, or to infer the significance of the regression from the results of the F-test described above. Each of these methods takes into account sample size.

Worked Example 9.5 Calculate the coefficient of determination for the tree height regression example presented above.

Solution For the tree height regression, $SS_Y = 113.151$ and $SS_E = 82.613$, so $R^2 = 1 - (82.613/113.151) = 0.27$, which indicates that the model accounts for 27% of the variability in the data. Whilst the regression is statistically significant (the amount of variance explained by the model is greater than that which would be expected by chance alone), the proportion of the variability in the dependent variable that can be explained (statistically) by changes in the independent variable is 27%; for the remaining 73% of the variability another explanation must be sought.

The same statistic can be calculated by using R. It is given under the name 'multiple R-squared' when a summary of the regression object is printed:

```
summary(treemodel)

Call:
lm(formula = height ~ girth)

Residuals:
    Min      1Q   Median      3Q      Max
-3.8359 -0.8441  0.0964  0.7539  3.0322

Coefficients:
             Estimate Std. Error  t value  Pr(>|t|)
(Intercept)   18.912      1.336    14.152  1.49e-14 ***
girth         12.656      3.868     3.272  0.00276  **
---
Signif. codes:  0 '***' 0.001 '**' 0.01 '*' 0.05 '.'
0.1 ' ' 1

Residual standard error: 1.689 on 29 degrees of freedom

Multiple R-squared: 0.2697, Adjusted R-squared: 0.2445

F-statistic: 10.71 on 1 and 29 DF,  p-value: 0.002758
```

Note that in the output from R the 'adjusted R-squared' value can be ignored in the present example, its use is relevant only in multiple regression where it accounts for the number of predictor variables that are being used, so that the R-squared value for models with a higher number of predictor variables is adjusted downwards.

9.8 Confidence intervals and hypothesis tests concerning regression parameters

9.8.1 Standard error of the regression parameters

Sometimes it is useful to know the precision with which the regression parameters m and c are known. It may also be of interest to test the hypothesis that the regression slope or intercept is equal to a particular given value. To fulfil either of these requirements the standard error of the regression parameters must be known. The standard error of a statistic is the standard deviation of its sampling distribution. In other words, the standard error measures the range of possible values of the statistic that would arise if we randomly resampled the data over and over again. In Chapter 3 the concept of a standard error was applied to the mean, but it can be calculated for any statistic, including the regression slope and intercept. If the standard error of the regression slope, SE_m, is large then we are less certain of its true value than if SE_m is small. The standard errors of the regression slope and intercept can be calculated from the mean square error which was defined above as: $MS_E = SS_E / (n-2)$ using the following formulae:

$$\text{'standard error of the slope'} \quad SE_m = \sqrt{\frac{MS_E}{SS_X}} \qquad (9.14)$$

$$\text{'standard error of the intercept'} \quad SE_c = \sqrt{MS_E\left[\frac{1}{n} + \frac{\bar{x}^2}{SS_X}\right]}.$$

$$(9.15)$$

9.8.2 Tests on the regression parameters

Having computed the standard error of the slope and the intercept, it is possible to test hypotheses about the regression parameters. Providing that the residuals are independent of each other and normally distributed (see Section 9.4), a null hypothesis about the regression slope or intercept can be constructed. For example, it may be sensible to test the null hypothesis H_0: $m = m_0$ where $m_0 = 1$ (i.e. there is a 1:1 relationship between the two variables) against the two-sided alternative H_1: $m \neq m_0$. In this case, the test statistic is:

$$t_{\text{calc}} = \frac{m - m_0}{\text{SE}_m} = \frac{m - m_0}{\sqrt{\dfrac{\text{MS}_E}{\text{SS}_X}}}, \tag{9.16}$$

which takes a t-distribution with $n - 2$ degrees of freedom. The null hypothesis, H_0, can be rejected at the α level of significance if $t_{\text{calc}} > t_{\alpha/2, n-2}$.

The special case where H_0: $m = 0$, which covers the situation that the fitted regression line is perfectly horizontal, is in fact equivalent to testing the significance of the regression using the F-test given in Section 9.6. It should be emphasized that failure to reject a null hypothesis that $m = 0$ does not imply that there is no relationship at all between the two variables, simply that there is no *linear* relationship between the variables.

Hypothesis tests for the intercept use the null hypothesis H_0: $c = c_0$ against the two-sided alternative H_1: $c \neq c_0$. The test statistic

$$t_{\text{calc}} = \frac{c - c_0}{\text{SE}_c} = \frac{c - c_0}{\sqrt{\text{MS}_E \left[\dfrac{1}{n} + \dfrac{\bar{x}^2}{\text{SS}_X} \right]}}, \tag{9.17}$$

takes a t-distribution with $n - 2$ degrees of freedom. Reject H_0 at the α level of significance if $t_{\text{calc}} > t_{\alpha/2, n-2}$.

Worked Example 9.6   Test the hypothesis that the slope of the tree height regression above is different from zero.

Solution   To test the null hypothesis that the regression slope is equal to zero against the two-sided alternative that it is not equal to zero requires a t-test, using the test statistic defined in Equation 9.10. The standard error of the regression slope is reported by R (under the label Std. Error) in the output from the `summary(treemodel)` command in Worked Example 9.5. In this case the standard error of the slope associated with the tree girth term in the model is 3.87. This can be calculated by hand from values of SS_X and MS_E which were computed in Worked Examples 9.2 and 9.3, respectively. With $SS_X = 0.19$ and $MS_E = 2.85$, $SE_m = 3.87$. The value of the test statistic is:

$$t_{calc} = \frac{12.66 - 0}{3.87} = 3.27, \tag{9.18}$$

which exceeds the critical value, $t_{\alpha/2,n-2} = 2.05$, leading to rejection of the null hypothesis and conclusion that the slope is different from zero. Note that t_{calc} is reported by R in the output from summary(treemodel), along with its p-value = 0.003. Further note that this p-value is identical to the p-value associated with the F-score in Worked Example 9.3. This correspondence highlights the equivalence noted earlier of the F-test based on components of variation and a t-test based on the regression slope.

9.8.3   Confidence intervals on the regression parameters

Under the assumption that the residuals are independent of each other and normally distributed, a $100(1-\alpha)\%$ confidence interval on the slope of the regression line is given by:

$$m \pm t_{\alpha/2,n-2} SE_m \tag{9.19}$$

Note that this expression is similar to the definition of the confidence interval for the mean given in Section 4.2. For the intercept, a similar approach gives:

$$c \pm t_{\alpha/2,n-2} SE_c \tag{9.20}$$

9.8.4 Confidence interval about the regression line

It is also useful to obtain a confidence interval associated with any prediction that might be made using the result of the regression. This measure is more complicated because the certainty with which the range of possible predictions can be known depends not just on the sample size and the variability of the underlying data, but also on how far the required prediction is from the values on which the regression estimates were developed. The error associated with a given prediction will be smaller if that prediction is close to the mean of the data upon which the regression relationships were developed than it will be if the required prediction is at a value which is far from the mean of the supporting data. Accordingly, a $100(1-\alpha)\%$ confidence interval on the mean response is given by:

$$\hat{y} \pm t_{\alpha/2,n-2} \sqrt{MS_E \left[\frac{1}{n} + \frac{(x_0 - \bar{x})^2}{SS_X} \right]} \tag{9.21}$$

Viewed graphically, the effect of making a prediction further from the mean of the data on which the regression analysis is based is obvious, and produced curved confidence bands (Figure 9.6).

9.9 Reduced major axis regression

The standard least-squares approach to linear regression introduced in Section 9.1 makes several assumptions (listed in Section 9.5). One of these is that the independent variable can be measured precisely and that the only source of error in the data lies in the dependent variable. In practice, this assumption is only very rarely true, perhaps in an experiment where the

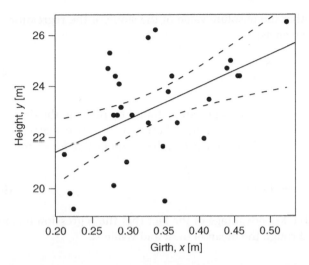

Figure 9.6 Confidence intervals (95%) around predictions of tree height.

values of the independent variable are fixed beforehand. Whilst in some geographical studies it is sensible to assume that the error in the dependent variable is substantially larger than error in the independent variable, in many studies this assumption does not hold. The most obvious cases include relations between two similar properties of the same entity, for example relating a river's width to its depth, or linking measures of poverty to measures of income. In each case, the amount of error in measuring each variable is similar and if we do not use a technique that accounts for the similarity explicitly we run the risk of calculating a misleading regression line. A specific account of the problem and an overview of possible solutions is given in Mark and Church (1977).

The solution that will be discussed here involves a modification to the linear least-squares approach in which the parameters of the regression line are estimated by minimizing the distance between the data points and the fitted line not just in the direction parallel to the vertical axis, as was assumed in Figure 9.2 but in a direction *perpendicular* to the fitted trend line itself. This procedure is called reduced major axis regression. The slope of such a line is straightforward to calculate and is simply the ratio of the standard deviations of the two variables, such that the absolute value of the slope of the regression line, m, is given as:

$$m = \frac{S_y}{S_x},\tag{9.22}$$

The sign of the slope can be found by plotting the data or by calculating the correlation between the two datasets (see Section 8.1). Having identified the correct value for m, the intercept, c, can be calculated as:

$$c = \bar{y} - m\bar{x},\tag{9.23}$$

which takes advantage of the fact that the regression line must pass through the mean of the x and y values.

9.10 Summary

The concepts introduced in the present section permit linear relations to be established between measured variables. The regression line can be defined in several different ways and, having defined the regression line, the analyst can partition the variability in the original data into an amount which is accounted for by the linear model and an amount which is not. If the model explains a substantial fraction of the original variability, then it is considered to be a good model and the given procedure for deciding whether the explanatory power of the model is more than would be expected by chance permits the analyst to decide whether or not the fitted model is statistically significant at a predetermined level of significance.

Exercises

1   A hydroecologist hypothesizes that for aquatic invertebrates species richness will decrease as rivers become more acidic. Using the data in Table 9.4 fit a linear model to explain variations in species richness in terms of pH.

Table 9.4 Species richness and pH for lowland stream reaches.

pH	Species richness
5.6	36
6.3	41
5.2	21
5.0	22
5.0	34
4.8	18
4.8	19
5.8	23
5.2	23
6.1	44
6.3	32
6.2	20
5.5	26
6.2	36
6.2	46
5.9	22
6.4	41
5.9	47
5.9	44
5.8	36
6.5	30
6.3	52
6.4	53
6.3	28
6.9	39
6.1	40
6.3	37
6.5	42
6.3	57
6.0	29
5.9	33
5.6	35
6.4	51
6.4	36

Source: Towsend *et al.*, 1983. Reproduced with permission of Wiley.

Table 9.5 Pollutant concentration away from a disused mine.

Distance (km)	Lead (Pb) concentration (ppm)
1.0	0.140
1.5	0.151
2.0	0.166
2.5	0.142
3.0	0.060
3.5	0.025
4.0	0.028
4.5	0.010
5.0	0.019
5.5	0.017
6.0	0.024
6.5	0.012
7.0	0.009
7.5	0.010
8.0	0.011
8.5	0.010
9.0	0.010
9.5	0.006
10.0	0.012

Source: Data from Selvey-Clinton, 2006.

2 Using the data in Table 8.3, produce a linear model to predict property prices as a function of distance from a city centre. Evaluate the significance of the regression and test whether the intercept and slope are significantly different from zero and one, respectively.

3 A researcher has reason to believe that there is a relationship between distance from an abandoned mine and pollutant concentration. Using the data in Table 9.5 test whether there is a linear relationship between the two datasets.

10

Spatial Statistics

STUDY OBJECTIVES

- Understand the differences between vector (points, lines, polygons) and raster (gridded) spatial data and know when each type might be used.
- Distinguish between different map projections used to display spatial data and know the typical circumstances in which each is likely to be used.
- Compute summary statistics for point datasets with coordinates and know the techniques used to visualize those datasets graphically.
- Understand the statistical procedures used to identify clusters in spatial datasets.
- Appreciate the geographical significance of spatial autocorrelation and know how to identify it in practical situations.

10.1 Spatial Data

10.1.1 Types of Spatial Data

The spatial data most commonly encountered in geographical studies can be divided into two broad categories: (i) *vector* data, which include points, lines and polygons; and (ii) *gridded* or raster data, which are specified on a regular grid.

Some examples of vector data include, locations of crimes committed (points), rain gauges (points), roads and railways (lines), geological faults (lines), parliamentary constituencies

Statistical Analysis of Geographical Data: An Introduction, First Edition.
Simon J. Dadson.
© 2017 John Wiley & Sons Ltd. Published 2017 by John Wiley & Sons Ltd.

(polygons), and lithological units (polygons). In some circumstances it is sufficient to record the presence or absence of a vector feature – for example the occurrence of a crime or the existence of a rain gauge. In other cases, a more complex set of attributes may be associated with a geographical feature for use in subsequent analysis. For example, it may be useful to record the date and nature of a crime rather than simply its location. This chapter is concerned with the analysis of spatial data in order to extract meaning from the spatial relationships between observations, and also to examine spatial patterns in measured values.

Raster (i.e. gridded) data also arise frequently in geographical investigations. Systematic sampling produces data that can be analysed using grid-based techniques. More recently the availability of remotely-sensed satellite data and the use of computer models has resulted in a large amount of gridded information which can be used in geographical analyses.

This chapter surveys some of the methods available to summarize and analyse spatial data. It also explores techniques to identify patterns amongst spatial data, including techniques to identify clusters within point datasets, and the presence of spatial relationships between mapped entities.

10.1.2 Spatial Data Structures

The distinguishing feature of spatial data is that each item within the dataset must have assigned to it some information about its location. This information is most commonly specified as a set of coordinates within a predetermined coordinate system. The most fundamental coordinate system used to specify locations on Earth (and other planets) is the spherical coordinate system in which locations are specified using longitude and latitude.

Longitude is the angular distance in the east–west direction. Lines marking points that have the same longitude are called meridians, and longitudes are conventionally given relative to a zero or prime meridian, which for historical reasons passes through Greenwich, UK. Latitude is the angular distance in the north–south direction. Lines of equal latitude are called parallels, the equator and the tropics of Cancer and Capricorn being the most well known. The equator has latitude 0° and latitudes are typically quoted north or south of the equator.

Latitudes and longitudes are typically quoted in degrees, minutes and seconds. Each degree is divided into 60 arc-minutes, and each minute into 60 arc-seconds, the prefix arc- being added in many situations to clarify that these are angular distances rather than units of time. A good rule of thumb is that, assuming that the radius of the Earth, r_e, is 6378 km at the equator, one degree corresponds to approximately $2\pi r_e/360 = 111$ km of linear distance. Therefore, one arc-minute of latitude or longitude is about 1.9 km, and one arc-second is about 31 m.

To take an example, the location of Oxford, UK is given as 51°45' N 1°15' W which tells us that it is 51 degrees and 45 arc-minutes north of the equator and 1 degree 15 arc-minutes west of Greenwich. To take another example, the Kenyan capital Nairobi has coordinates 1°17' S 36°49' E.

In situations where the units of degrees, arc-minutes and arc-seconds might be considered cumbersome, the alternative system of decimal degrees can be used. Here, the arc-minutes and arc-seconds are converted to a decimal fraction of a degree. By convention, north and east are taken to be positive and west and south are negative. For example, to two decimal places, Oxford (51°45' N 1°1'5' W) becomes 51.75°, −1.25° and Nairobi (1°17' S 36°49' E) becomes −1.28, 36.82.

Worked Example 10.1 Produce a map showing the global distribution of all cities that have populations greater than 5 million. Represent each city with a symbol whose size is in proportion to the city's population.

Solution To address this challenge, we need first to plot a base map showing the outlines of countries of the world and then to plot some data on world cities on top of the base map. Fortunately, R has built in datasets to handle both these tasks. The relevant functions are found in the packages listed in Table 10.1.

The first task is to load these packages. It may be necessary to install them first using R's package installer (see Appendix A for instructions on how to install new packages).

A basic map of the world can be plotted using the command:

```
map("worldHires", xlim=c(-180,180),
    ylim=c(-60,90), col="lightgray", fill=TRUE)
```

which chooses the basemap called "worldHires". As the name suggests, this is a high resolution map of the world containing

Table 10.1 R packages for plotting and analysing spatial data.

Package	Function
sp	Provides new data structures to enables coordinate information to be attached to dataframes, and procedures for spatial analysis
maps	Provides the facility to plot maps
mapdata	Provides high-resolution base maps for use with the 'maps' package
mapproj	Provides R with the ability to transform coordinates between coordinate systems using different map projections

country polygons. The extent of the displayed map can be specified by using the xlim and ylim arguments. Note that in this case xlim has been set to vary from −180 to 180 which covers the full range of longitudes, but ylim has been restricted to start at −60 (i.e. 60°S) in order to exclude Antarctica. To add axes to the map, and to overlay a grid of latitudes and longitudes, use:

```
map.axes()
```

```
grid()
```

Note that the default method for displaying map data is to plot latitudes and longitudes in equally spaced increments. This is the most basic map projection and is referred to as 'geographic', 'rectangular' or 'plate carrée (= square plate)' projection. More sophisticated map projections that minimize different types of distortion are discussed in Section 10.1.3, below.

It remains to plot the data themselves. R contains a built-in dataset that has information on the population of many cities in the world together with their latitude and longitudes. We can access and examine a sample of that dataset using:

```
data(world.cities)
```

```
head(world.cities)
```

For the present application, it is not necessary to plot all of the cities. A subsample of those cities with population greater than 5 million people is needed. The standard subsetting operations in R can be used:

```
cities_5m <- world.cities[world.
  cities$pop>5000000,]
```

A quick look using `summary(cities_5m)` indicates that there are 27 cities that meet the criterion. To plot the cities on the map, we need to convert them into a spatial object by telling R that the variables 'lat' and 'long' information correspond to geographical latitude and longitude.

```
coordinates(cities_5m) = c('long','lat')
```

It is also necessary to supply R with the correct coordinate information, namely that these are data given as latitudes and longitudes on the WGS84 coordinate system. This is achieved by specifying the coordinate system in PROJ.4 format. PROJ.4 format is a standardized system for representing map projections (Evenden, 2003).

```
proj4string(cities_5m) <- CRS('+proj=longlat
  +datum=WGS84')
```

With that information, the positions of the cities can be plotted by using the points command:

```
points(cities_5m, pch=21,
  cex=(cities_5m$pop)/5000000, col="black",
  bg="white")
```

where the symbology has been set to `pch = 21` so that symbol sizes are scaled by population (i.e. a city with population equal to 5 million people is plotted with a `cex` value of 1.0). The resulting map is shown in Figure 10.1.

10.1.3 Map Projections

Map projections are used to convert angular coordinates measured on a sphere into rectangular coordinates that can be plotted on a paper map or a computer screen. It is inevitable that map projections will introduce some distortion. It would be impossible, for example, to peel an orange and shape the peel into a flat rectangle without tearing or squashing it in some way. The question is how to minimize the level of distortion.

There are two important stages involved in transforming spherical coordinates into coordinates on a plane. The first is to

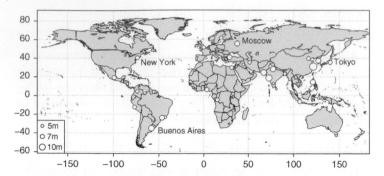

Figure 10.1 Map showing global cities with population greater than five million people. Source: Data from R built-in dataset.

approximate the shape of the Earth using a reference ellipsoid. The standard World Geodetic System 1984 (WGS84) reference ellipsoid has an equatorial radius of approximately 6378 km and a polar radius of 6357 km (NIMA, 1997). This ellipsoid is used in most modern global mapping and is the standard for use when finding coordinates using the satellite-based Global Positioning System (GPS).

The second step in producing a projected map is to select a map projection. There are many different map projections available, for use in different situations. Several of the most frequently used are described in Figure 10.2 (see Evenden, 2003, for a more detailed account). The major groups are cylindrical and conical and they derive their names from the method of construction. For example, the Mercator projection, which is named after Gerardus Mercator (1512–1594), and is one of the best-known projections, is a cylindrical projection. To construct a cylindrical projection, imagine a cylinder of paper wrapped around the globe so that it touches the surface of the sphere only along the equator. The locations of points on the sphere are then projected onto the paper as though a light source were placed at the centre of the sphere. Inevitably distortions occur in this process, so that points where the cylinder of paper touches or is close to the sphere will be distorted less than those further away. In the case of the Mercator projection, points near to the equator are distorted less than those at the poles. This property makes the Mercator projection misleading

Mercator projection

- Lines of constant bearing appear straight.
- Extreme distortion at high latitudes (shaded)

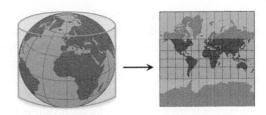

Transverse mercator projection

- Distortion increases away from central meridian
- e.g., UTM, Great Britain Ordnance Survey grid

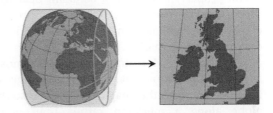

Polar stereographic projection

- Distortion increases with distance from pole
- e.g., Universal Polar Stereographic Projection

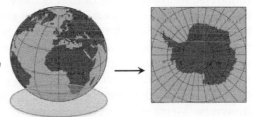

Lambert conformal conical projection

- Conical projection used in mid-latitudes
- Appropriate for maps aligned E-W

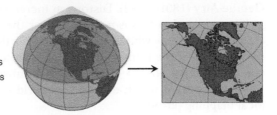

Figure 10.2 Map projections and their properties

when comparing areas at different latitudes, and gives rise to the widely known deficiency with world maps produced using the Mercator projection which is that Greenland appears to be larger than Africa when it is, in fact, 13 times smaller. This is generally speaking not a deliberate deception on the part of those who use maps produced using the Mercator projection, but comes from its widespread adoption in the 16th century for use in marine navigation where the Mercator projection has the useful property that lines of constant bearing appear straight on the map. In general use, the Mercator projection is a sensible choice only when mapping regions close to the equator where distortion is minimal.

The transverse Mercator projection is almost identical to the Mercator projection but, as its name suggests, this method involves wrapping the imaginary cylinder so that it touches the globe along a meridian rather than along the equator (Figure 10.2). Any central meridian can be chosen to create a transverse Mercator projection. It is therefore along this central meridian that the distortion is zero; away from it the distortion increases in the same way as with the Mercator projection. Transverse Mercator projections are commonly used when mapping regions which have north–south alignment, including the United Kingdom, and the British National Grid defined by the Ordnance Survey is in fact based on a transverse Mercator projection with 2° W 49° N as its origin, and an ellipsoid defined by British Astronomer Royal Sir George Airy (1801–1892). Distortion increases away from the 2° W meridian but this is negligible over the landmass of the United Kingdom, which is oriented north–south. Other notable uses of the transverse Mercator projection are in the construction by NATO of the UTM (Universal Transverse Mercator) system which divides the world into 60 zones for local mapping purposes.

Another important projection, which is especially suited to the polar regions is the polar stereographic projection. This projection is constructed by placing a plane tangential to the Earth at one pole and projecting points from a source at the other pole towards the plane (Figure 10.2). In this projection, meridians are projected as straight lines radiating outwards from the chosen pole whilst parallels form concentric circles. The UTM system does not cover the poles, and so for latitudes above 84° N or below 80° S the Universal Polar Stereographic (UPS) projection is often used instead (Snyder, 1987).

Conical projections are another commonly used class of map projections. They are constructed by imagining a cone of paper wrapped around the Earth, rather than a cylinder. The Lambert conformal conic projection is an important example of this projection. It is based upon an imaginary cone wrapped around a chosen parallel. Points on the globe are then mapped onto the surface of the cone. This projection is useful in mid-latitude locations and is commonly used for geographical regions which are aligned east–west. The continental USA and Europe are particularly suited to display using the Lambert conformal conical projection. To illustrate the differences, Figure 10.3 shows a comparison between the continent of Europe projected using Mercator versus Lambert conformal conic.

Worked Example 10.2 Plot the locations of the 30 most populous cities in the UK on a basemap displayed in the OS GB National Grid coordinate system.

(a)   (b)

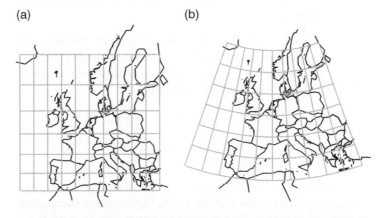

Figure 10.3 Comparison of Europe plotted with (a) Mercator projection and (b) Lambert conformal conic (ETRS89-LCC) with standard parallels at 35° N and 65° N. Note the relative sizes of different countries, especially at high latitudes.

Solution This example involves map projection and so in addition to the libraries loaded in Worked Example 10.1, it is necessary also to load the mapproj and rgdal libraries:

```
library(mapproj)
library(rgdal)
```

R's built-in selection functions can be used to extract the top 30 cities in the UK:

```
data(world.cities)
uk.cities <- world.cities[world.
  cities$country.etc=='UK',]
uk.cities <- uk.cities[order(-uk.cities[,3]),]
uk.cities <- uk.cities[1:30,]
```

These commands are straightforward, but note that uk. cities[order(-uk.cities[,3]),] is R's way of saying sort the cities in order of descending population (descending hence the minus sign, and column 3 contains the data on population).

The next two tasks involve the assignment of spatial coordinates and projection information.

```
coordinates(uk.cities) = c('long', 'lat')
proj4string(uk.cities) <- CRS('+proj=longlat
  +datum=WGS84')
```

These steps are followed by reprojection from latitudes and longitudes into coordinates on the GB National Grid. First, it is necessary to define the new projection. The projection is specified in PROJ.4 format using the command:

```
gb_grid <- CRS('+proj=tmerc +lat_0=49
  +lon_0=-2 +k=0.9996012717 +x_0=400000
  +y_0=-100000 +ellps=airy +datum=OSGB36
  +units=m +no_defs')
```

which tells R that the GB National Grid is a transverse Mercator projection with its origin at 2° W, 49° N. Various other parameters are required, including the ellipsoid and datum (see Table 10.2).

To transform the gb_cities dataset which was created above into the GB National Grid coordinate system, type:

Table 10.2 Parameters of the Ordnance Survey Great Britain projection (OSGB36).

Parameter	Value
Projection	Transverse Mercator
Origin longitude	2° W
Origin latitude	49° N
False easting	400 km west of true origin
False northing	100 km north of true origin
Scale factor	0.9996012717
Ellipsoid	Airy

```
uk.cities_osgb <- spTransform(uk.cities,
  gb_grid)
```

A comparison of the resulting coordinates with the original ones indicates that the transformation has been successful (only the top five cities are listed here):

```
> uk.cities[1:5,]
          coordinates  name country.etc      pop capital
21344  (-0.1, 51.52)  London       UK  7489022       1
4503   (-1.91, 52.48) Birmingham   UK   986969       0
12619  (-4.27, 55.87) Glasgow      UK   607192       0
21165  (-2.99, 53.42) Liverpool    UK   468584       0
20527  (-1.55, 53.81) Leeds        UK   457875       0

> uk.cities_osgb[1:5,]

                 coordinates  name country.etc pop capital
21344   (531921.6, 181831) London      UK 7489022       1
4503    (406209.5, 286901) Birmingham  UK  986969       0
12619   (258058.2, 666422.3) Glasgow   UK  607192       0
21165   (334303.7, 391923.4) Liverpool UK  468584       0
20527 (429729.8, 434950.5) Leeds       UK  457875       0
```

Note that R tracks the full coordinate information in units of metres from the origin of the OS National Grid system (which is located west of the Scilly Isles).

The next task is to reproject the base map. This can be achieved by extracting the lines that make up the map and reprojecting them as above:

```
uk_coast_poly <- map("worldHires",
  c("UK:Great Britain", "UK:Scotland",
  "UK:Northern Ireland", "UK:Isle of
  Sheppey"), fill=TRUE, col="transparent",
  plot=FALSE)

IDs <- sapply(strsplit(uk_coast_poly$names,
  ":"), function(x) x[1])

uk_coast_poly_sp <- map2SpatialPolygons(uk_
  coast_poly, IDs=IDs,
  proj4string=CRS("+proj=longlat
  +datum=WGS84"))

uk_coast_poly_osgb <- spTransform(uk_coast_
  poly_sp, gb_grid)
```

Now that the points and the basemap are both in the desired projection, it remains to create the plot:

```
plot(uk_coast_poly_osgb, xlab="km",
  xaxt="n")

xpos <- seq(0, 8e5, by=2e5)

ypos <- seq(0, 12e5, by=2e5)

axis(1, at=xpos, labels=sprintf("%.0f",
  xpos/1000))

axis(2, at=ypos, labels=sprintf("%.0f",
  ypos/1000))

title(xlab="Easting [km]", ylab="Northing
  [km]")
```

Note that the axes have been customized to show units in kilometres whilst keeping the underlying dataset in metres.

Finally, to display the points themselves, type:

```
points(uk.cities_osgb, pch=16)
```

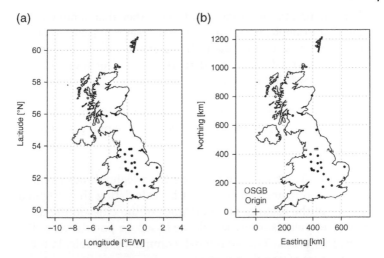

Figure 10.4 Plot showing the locations of the top 30 cities in the UK by population: (a) in latitude-longitude coordinates; and (b) projected onto GB National Grid coordinates using R.

The resulting plot is shown in Figure 10.4b, with the original latitude–longitude equivalent beside it as Figure 10.4a. Note the location of the origin of the OS National Grid which has been marked in Figure 10.4b too.

10.2 Summarizing Spatial Data

10.2.1 Mean Centre

The mean centre provides a good quantitative measure of central tendency to use with spatial datasets. The mean centre is calculated by finding the arithmetic mean of the x-coordinates and the mean of the y-coordinates. In order to calculate the mean centre of a group of points it is first necessary to project the points into a rectangular coordinate system. Together the means of the x- and y-coordinates form the coordinates of the mean centre.

10.2.2 Weighted Mean Centre

Situations in which the points do not warrant equal weighting require the use of the weighted mean centre where a weighted average of the coordinate values is found. This approach is useful when the points represent quantities rather than simply locations. For example, if each point in a dataset represents a settlement within a region then the mean centre weighted by population would give a good representation of the centre of population within the region.

Worked Example 10.3 Calculate the position of the mean centre of the UK population and add it to the plot produced in Worked Example 10.2

Solution The trick here is to use the weighted mean centre, whose coordinates can be found by calculating the weighted average of the eastings and northings of the cities in the database. The weights to use should be based on population of the individual cities. The additional commands to calculate the mean centre of population are:

```
mc_easting <- weighted.mean(uk.cities_osgb@
  coords[,1], uk.cities$pop)

mc_northing <- weighted.mean(uk.cities_osgb@
  coords[,2], uk.cities$pop)

mc <- cbind(mc_easting, mc_northing)

points(mc, pch=21, cex=3, bg="gray", lwd=2)
```

The mean centre of the UK population is at 430735 E, 308263 N, which corresponds to a location near Ashby de la Zouch in Leicestershire (Figure 10.4).

10.2.3 Density Estimation

The mean centre provides a useful summary but it does not tell us about the spatial distribution of points across the region of interest. A straightforward method for estimating the spatial distribution of a set of points is to use density estimation. This technique uses a moving window (which is called a kernel) to count the number of points that lie within a unit area across the region of interest. In the case of a quantity such as population it is also possible to weight the density estimate by the population of the settlements being analysed in order to reach an estimate of population density rather than settlement density, as illustrated in Figure 10.5. Many further

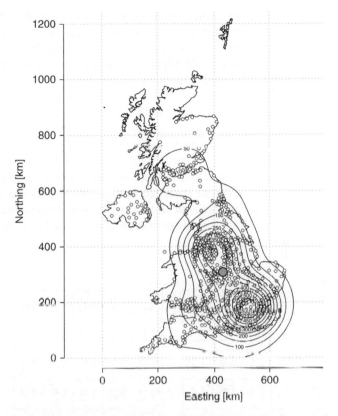

Figure 10.5 Geographical distribution of the UK population. Open circles indicate individual towns. The filled grey circle is the mean centre. Solid lines are contours of population density in persons per square kilometre calculated using a 60 km averaging kernel.

aspects of density estimation are covered in more advanced texts (e.g., Diggle, 2002).

10.3 Identifying Clusters

10.3.1 Quadrat Test

Thus far, we have been concerned the production of graphical and numerical summaries of point data. Many studies call for inferential statistics concerning the presence or absence of clusters in point data. Statistically, we want to pose the question: are the point data spread uniformly in space or are they clustered? In other words, does the distribution of the points that we have observed differ from what we would have expected had the points been distributed randomly across the study area. Some of the ways in which a point pattern could hypothetically be distributed are shown in Figure 10.6. Figure 10.6a shows points that are completely spatially random (CSR). There are two alternative possibilities. The first is that the points are more spread-out than would be expected had they resulted from a completely spatially random process (Figure 10.6b). The second possibility is that the points tend to occur in clusters rather than at random (e.g. Figure 10.6c). It can be seen that even when the generating process is CSR, the presence of some clusters is to be expected; they are the geographical equivalent of getting several heads in a row when tossing a coin.

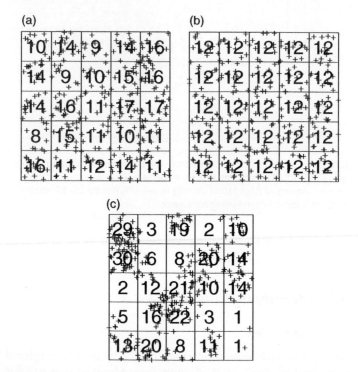

Figure 10.6 Possible distributions of points across a region: (a) completely spatially random; (b) dispersed; and (c) clustered. Quadrat outlines are drawn with numbers indicating the number of points that lie within each quadrat.

A statistical test to work out whether the observed amount of clustering is more than would be expected at random can be devised by marking out a set of quadrats which span the study region. The number of points in each quadrat is counted and the counts are compared with the number that would be expected to have occurred in each quadrat if the generating process were spatially random. This procedure, which is illustrated in Figure 10.6, is called the quadrat test. It is similar to the Chi-squared test that was introduced in Chapter 6 and the mathematics of the test are identical. The quadrat test can be performed in R using the library spatstat, which contains the function quadrat.test(). The results from applying this test to the pattern in Figure 10.6a are:

```
> quadrat.test(pp0)

    Chi-squared test of CSR using quadrat counts

data: pp0

X-squared = 14.4361, df = 24, p-value = 0.1276

alternative hypothesis: two.sided

Quadrats: 5 by 5 grid of tiles
```

The p-value returned by the test is quite high (much greater than 0.05) and indicates that there is a high probability that the data in Figure 10.6a could have resulted from a spatially random process (which is good because that is in fact how they were generated). By contrast, application of the quadrat test to the data in Figure 10.6c results in the following output, which indicates that there is only a very small probability that the data arose from a spatially random process. The resulting value of Chi-squared is very high, suggesting that there are some quadrats that have more than expected number of points, and some quadrats that have fewer:

```
> quadrat.test(pp2)

    Chi-squared test of CSR using quadrat counts

data: pp2

X-squared = 145.1667, df = 24, p-value <
    2.2e-16

alternative hypothesis: two.sided

Quadrats: 5 by 5 grid of tiles
```

Finally, the result from applying the quadrat test to Figure 10.5b shows that there is a low probability that these data arose by chance too:

```
> quadrat.test(pp1)

 Chi-squared test of CSR using quadrat counts

data: pp1

X-squared = 0, df = 24, p-value < 2.2e-16

alternative hypothesis: two.sided

Quadrats: 5 by 5 grid of tiles
```

In this case, however, the data are too structured to be considered random. There is an identical number of points in each quadrat: a situation that would be highly unlikely to occur as the result of a spatially random process.

10.3.2 Nearest Neighbour Statistics

The quadrat test is the simplest test that can be used to diagnose patterns in spatial point datasets. It is robust and easy to understand if one knows about Chi-squared already. However, this test suffers from the problem that the result can in some cases depend on the properties of the chosen quadrats, especially their size and shape. A more comprehensive approach, which is independent of the choice of quadrat size, is to look instead at the distance between each point and its nearest neighbour. In the event that the data are clustered, the nearest neighbours will on average be close by. Conversely, if the data are structured then the nearest neighbour distances will be more uniformly distributed. This test is beyond the scope of the present book, but it is described in detail by Cressie (1993).

10.4   Interpolation and Plotting Contour Maps

Interpolation is the process whereby measurements of a quantity at a set of points distributed across a region are generalized so that an estimate of the value of the quantity can be provided at locations within the region where no measurements are

available. Many techniques exist for interpolating point datasets, and the resulting end product can take the form of a contour map, in which isolines or contours join points that share a
particular data value, or a gridded dataset in which regular estimates of the quantity in question are given on a regular grid.

Examples of contour maps in common use include topo-
graphic maps used in hydrology and geomorphology, and air
pressure maps used in meteorology. However, any measure that
varies in space can be summarized using a contour map, providing that there are not too many discontinuous jumps in the value
of the plotted quantity. To produce a contour map by hand is
relatively easy: simply draw lines that join points of equal value,
using interpolation in cases where points have values that do not
coincide with the desired contour interval (e.g. the 10 m contour
should pass half way between two points measured at 5 and
15 m elevation).

10.5 Spatial Relationships

10.5.1 Spatial Autocorrelation

One of the most important questions that can be posed when
dealing with spatial data is whether a phenomenon occurs at
random across a geographical region, or whether it is clustered or dispersed. These terms were encountered in
Section 10.2 when studying point datasets and a number of
techniques were proposed for quantifying the effect of proximity on relationships between spatial entities. Many types of
geographical data are measured in regions that are represented as polygons (e.g. country-level data, electoral statistics). When working with spatial data measured in polygonal
regions, an alternative suite of techniques is available to
investigate spatial autocorrelation.

Whilst spatial proximity between points can be measured easily using their distance apart, the spatial proximity of contiguous
polygons must be defined by looking at which polygons share
common edges. The first step in any analysis of autocorrelation
amongst polygonal datasets is to establish which polygons are
connected to each other. It is possible to work this out by hand,
although the process can be tedious if there are many connected

Figure 10.7 Polygons showing London boroughs. Filled circles show the centres of each borough; solid grey lines connect boroughs which share a boundary. Note that the River Thames passes through the centre of the city but has been ignored for the purposes of evaluating shared boundaries. Produced using data from the Office for National Statistics licensed under the Open Government Licence v.3.0. © Crown copyright and database rights (2015) Ordnance Survey 100019153. http://data. london.gov.uk/datastore/package/statistical-gis-boundary-files-london.

areas or if their geometries are complicated. Fortunately, automated methods are available to work out which polygons share a common boundary using software such as R (Figure 10.7, and see Worked Example 10.4).

10.5.2   Join Counts

The simplest way to measure autocorrelation with areal datasets uses join counts. However, it is applicable only to data that can be expressed in binary form (i.e. where values are in one of only two categories). This restriction may at first appear onerous, but there is a large range of geographical examples for which binary data can provide meaningful insight. One example includes the electoral dataset depicted in Figure 10.8, which shows voters' preferences for Labour versus Conservative candidates in the London Mayoral Election of 2012. Similar examples include other electoral datasets, and many other kinds of binary data on,

Figure 10.8 Voting patterns in the 2012 London Mayoral election. Polygons represent London boroughs, shaded to indicate whether the majority of first preference votes was for the Labour Party candidate (shown in light grey) or the Conservative Party candidate (shown in dark grey). Produced using data from the Office for National Statistics licensed under the Open Government Licence v.3.0. © Crown copyright and database rights (2015) Ordnance Survey 100019153.

for example, land use (rural–urban), crime (above or below average), or even the mapped residuals from regression analyses which can be coded as positive or negative. The statistical basis for the joint count technique is that, under the null hypothesis, the phenomenon of interest is distributed spatially at random. If this is the case, there is a known probability that a light grey area will adjoin a dark grey area. If a higher number of joins between light grey and dark grey areas is observed than would be expected at random then the analyst is justified in concluding that there is a spatial pattern amongst the data.

The mathematical rationale for such a test rests on the fact that if the light grey and dark grey areas are mixed at random, the expected number of light–dark joins, E_{GD}, can be calculated as:

$$E_{GD} = \frac{2JGD}{n(n-1)}, \tag{10.1}$$

where J is the total number of joins and G and D are the numbers of light and dark grey areas, respectively. The observed number of light–dark joins, O_{GD}, is found by counting them on the map, and the test statistic is obtained by subtracting the expected number of light–dark joins from the number observed and normalizing by the standard error of the estimate, σ_{GD}, which itself is calculated according to Cliff and Ord (1981):

$$\sigma_{GD} = \sqrt{E_{GD} + \frac{\sum L(L-1)GD}{n(n-1)} + \frac{4\left[J(J-1) - \sum L(L-1)\right]G(G-1)D(D-1)}{n(n-1)(n-2)(n-3)} - E_{GD}^2}$$

(10.2)

The term $\sum L(L-1)$ in the above equation is calculated so that for each region, L is the number of joins between that area and its immediate neighbours.

Combining these calculations yields the test statistic:

$$z = \frac{O_{GD} - E_{GD}}{\sigma_{GD}}$$

(10.3)

which follows a standard normal distribution (i.e. a normal distribution with mean of zero and unit standard deviation). The p-value for the test can be obtained by calculating the probability that the difference between observed and expected values is greater than the calculated value. The calculations can be handled automatically in R, as illustrated in Worked Example 10.4.

Worked Example 10.4   Using join count statistics, find out whether the voting patterns in the London Mayoral Election were spatially random or whether they contained a significant level of clustering.

Solution   The initial step in this analysis is to read the appropriate datasets into memory. The first dataset contains a map of the polygons that represent the London Boroughs. These data are stored in shapefile format, and can be accessed using the `getinfo.shape` and `readShapePoly` functions within the maptools package:

```
library(maptools)
```

```
getinfo.shape("wex10_4.shp")
```

The getinfo.shape() function reports that the file contains 33 polygons representing the 32 Boroughs together with the City of London, which is technically not a Borough. The shapefile can be read in using:

```
london <- readShapePoly("wex10_4.shp")
```

To plot the shapefile, the plot command can be used:

```
plot(london)
```

However, it should be noted that whilst the shapefile gives the location and geometry of the Borough boundaries, it contains relatively little information on the inhabitants of each Borough. Extra information on geographical properties of interest is provided in Table 10.3.

These data are contained in the spreadsheet london boroughs.csv and can be read into R as follows:

```
london_boroughs <- read.csv("wex10_4.csv")
```

Merging the two datasets requires R to find rows that share a common identifier and produce a new dataset that combines the two datasets. In the present case we want R to match the names of the boroughs in order to assign values to the new data table (i.e. the merge field is "NAME").

```
london_merge <- merge(london, london_bor-
   oughs, by="NAME")
```

The next step in the analysis is to calculate which of the polygons are neighbours. The function to make this calculation is poly2nb. It is built in to the spdep package which is designed to investigate spatial dependence within datasets. To perform the calculation use:

```
london_merge.nb <- poly2nb(london_merge,
   snap=500)
```

Table 10.3 Data for London Boroughs (2013/14 figures).

Borough	Median house price (£)	Unemployment (%)	Crime rate (per thousand population)	Ambulance incidents (per hundred population)	Mayor
City of London	595 000	–	–	81	CON
Barking and Dagenham	180 000	13	83	14	LAB
Barnet	346 000	5	61	12	CON
Bexley	225 000	9	50	12	CON
Brent	350 000	9	77	13	LAB
Bromley	295 650	5	61	11	CON
Camden	575 000	7	124	16	LAB
Croydon	240 000	9	76	14	CON
Ealing	327 000	11	75	12	LAB
Enfield	250 000	7	70	13	LAB
Greenwich	265 000	11	74	13	LAB
Hackney	375 000	11	101	12	LAB
Hammersmith and Fulham	570 000	6	106	13	CON
Haringey	339 875	9	84	13	LAB
Harrow	325 000	9	50	10	CON
Havering	231 000	7	62	13	CON
Hillingdon	265 000	8	67	16	LAB
Hounslow	280 000	8	75	12	LAB
Islington	462 726	8	114	15	LAB
Kensington and Chelsea	980 000	9	115	13	CON
Kingston upon Thames	335 000	6	55	12	CON
Lambeth	345 000	8	104	14	LAB
Lewisham	258 000	9	78	13	LAB
Merton	322 000	6	57	11	CON
Newham	230 000	11	91	13	LAB
Redbridge	275 000	9	69	12	LAB

(Continued)

Table 10.3 (Continued)

Borough	Median house price (£)	Unemployment (%)	Crime rate (per thousand population)	Ambulance incidents (per hundred population)	Mayor
Richmond upon Thames	475 000	5	55	10	CON
Southwark	347 500	11	104	15	LAB
Sutton	250 000	6	53	12	CON
Tower Hamlets	320 000	13	99	12	LAB
Waltham Forest	263 000	7	83	12	LAB
Wandsworth	461 125	7	71	11	CON
Westminster	730 000	4	238	22	CON

Source: Produced using data from the Office for National Statistics licensed under the Open Government Licence v.3.0. data.london.gov.uk.

which takes each polygon in turn and identifies polygons that share a boundary as 'neighbours'. The argument 'snap' allows us to specify that we can tolerate a margin for error in the shared boundary of 500 m; this facility is useful in the present context because it has the effect of including boroughs which are neighbours across the river. The resulting data structure contains a list of neighbours, as follows:

```
> London_merge.nb
Neighbour list object:
Number of regions: 33
Number of nonzero links: 166
Percentage nonzero weights: 15.24334
Average number of links: 5.030303
```

To compute join count statistics, the command joincount is used. The requested variable here is Mayor, within the London_merge dataset.

```
joincount.test(london_merge$Mayor,
  nb2listw(london_merge.nb))
```

The resulting output is as follows:

```
Join count test under nonfree sampling
data:   london_merge$Mayor
weights: nb2listw(london_merge.nb)
Std. deviate for CON = 1.6226, p-value =
  0.05234
alternative hypothesis: greater
sample estimates:
Same colour statistic   Expectation Variance
              4.0559524    3.2812500  0.2279628


        Join count test under nonfree sampling
data:   london_merge$Mayor
weights: nb2listw(london_merge.nb)
Std. deviate for LAB = 2.4998, p-value =
  0.006212
alternative hypothesis: greater
sample estimates:
Same colour statistic   Expectation  Variance
              6.0154762    4.7812500   0.2437609
```

The test statistic is higher than would be expected with a random distribution, which indicates a clustered pattern. The *p*-value gives the probability of measuring a test statistic this high or higher under the assumption of a spatially random distribution. In this case, the *p*-value is less than 0.05 and so the test therefore indicates a statistically significant level of clustering in this dataset, at the 5% significance level.

Other more advanced techniques for evaluating whether spatial patterns exist in data sets are beyond the scope of this introductory treatment. For example, Moran's *I*-statistic is particularly useful when looking at interval scale data. See Cressie, 1993, for full details.

Exercises

1 Use a GPS to find your current latitude and longitude. With the help of R's `spTransform` command, and other functions described in Section 10.1.3 convert that position into coordinates on the Ordnance Survey National Grid.

2 One of R's built-in datasets named `meuse` contains information on heavy metal contamination from soil samples collected on the floodplain of the River Meuse in the Netherlands. Load the dataset using the command `data(meuse)` and, with the aid of the functions described in Section 10.1.2 produce a map of the data showing concentrations of the various heavy metals in different colours.

3 The dataset `redwood` gives the coordinates of 62 California redwood seedlings and saplings. Load the data into R using the command `data(redwood)` and, by using the quadrat test described in Section 10.3.1, determine whether the spatial distribution of the trees is random. Consider the assumptions and limitations of your analysis.

11

Time series analysis

STUDY OBJECTIVES

- Recognize and understand the defining properties of time series data.
- Understand how R handles time series data.
- Know how to plot time series data using R and be aware of the factors to consider when doing so.
- Understand how to diagnose common modes of variability in time series analysis including trends and seasonality.
- Appreciate the importance of autocorrelation in time series data and understand how the correlogram can be used to represent autocorrelation.

11.1 Time series in geographical research

Time series data are common in geographical applications. The properties of a time series were introduced in Chapter 2 by using the example of the monthly measurements of carbon dioxide (CO_2) made at the Mauna Loa Observatory in Hawaii (see Figure 11.1 which shows the time series again). Put simply, a time series is any dataset that contains observations that are arranged sequentially in time. Examples of geographical quantities that might be measured in a time series include temperature, wind speed, humidity, rates of species extinction, and rates of cliff erosion over time. In human geography, examples of time

Statistical Analysis of Geographical Data: An Introduction, First Edition.
Simon J. Dadson.
© 2017 John Wiley & Sons Ltd. Published 2017 by John Wiley & Sons Ltd.

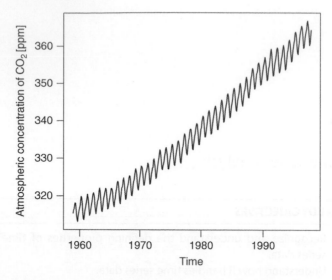

Figure 11.1 Time series of Mauna Loa atmospheric CO_2 concentrations. Note that this dataset is built into R and can be plotted using `plot(co2)`. Source: Keeling, C. D. and Whorf, T. P., Scripps Institution of Oceanography (SIO), University of California, La Jolla, California USA 92093-0220.

series may include economic data on gross domestic product (GDP) in successive years, demographic data on population growth and rates of migration, and data on how life expectancy has changed over time. The aim of this chapter is to provide some introductory tools and techniques that can be used to analyse time series in order determine how different systems behave over timescales of various durations. Special techniques to investigate trends and seasonal cycles are presented, and some of the particular problems that often arise in the analysis of time series are discussed.

11.2 Analysing time series

11.2.1 Describing time series: definitions

Several types of time series can be defined at the outset: *discrete* time series are sampled only at particular times – these intervals are usually regularly spaced although they may not always be so. Some discrete time series may contain sampled data that have been measured instantaneously in time (e.g. temperature); other datasets may contain aggregate or accumulated data that represent the total measured quantity over a specific time interval. Examples of the latter include rain gauge data, which are often provided as daily or monthly rainfall accumulations.

By contrast, *continuous* time series are recorded continuously. Truly continuous time series are rare in the era of digital instruments, but analogue chart recorders used in some tipping-bucket recording rain gauges and seismometers provide an example of a commonly collected form of continuous time series.

11.2.2 Plotting time series

Working with time series data in R is made easier through the use of R's time series data structure. The main benefit of using the time series data structure is that in addition to the observations themselves, R keeps track of the start and end date of the time series, as well as its time step. This information is grouped together and stored in memory as a time series 'object'. For the user, the big advantage is that many of the standard R methods and functions such as plot, mean, std, have been adapted to work with time series. These modified functions simplify the number of choices that must be made when plotting time series and provide finer control over the output. For example, when told to plot a time series object R will interrogate the data contained within the object and make sensible default choices about the time increments to use on the horizontal axis which the user can then tailor to his or her specific requirements.

Reading time series into R is a straightforward process, but it involves two distinct stages. First, the raw data must be read into memory. Then a time series object is created by adding information on the start, end, and time increment.

Worked Example 11.1 Read the monthly rainfall data for Oxford, UK, 1961–1990 contained in Table 11.1 into R and create a time series object. Plot the data making sensible choices for the axis scales and symbology.

Table 11.1 Mean monthly temperature data (°C) for Oxford, UK, 1961–1990.

	Jan	Feb	Mar	Apr	May	Jun	Jul	Aug	Sep	Oct	Nov	Dec
1961	4.2	7.6	8.7	10.7	11.8	15.6	16.5	16.6	16.0	11.3	6.2	2.5
1962	4.6	4.6	3.3	8.3	10.9	14.6	16.0	15.5	13.3	10.8	5.6	1.3
1963	-3.0	-0.7	6.5	9.5	11.3	15.5	15.9	15.0	13.7	11.6	8.9	2.5
1964	3.4	5.0	4.7	9.2	14.0	14.7	17.4	16.8	15.1	9.1	7.8	3.8
1965	3.8	3.1	6.0	8.8	12.4	15.0	14.8	15.6	13.0	11.1	5.0	5.3
1966	2.9	6.8	7.0	8.1	11.8	16.1	15.8	15.5	14.4	11.1	5.6	6.1
1967	4.9	5.9	7.7	8.4	11.6	14.7	17.9	16.5	14.3	11.5	6.0	4.6
1968	4.7	2.3	7.2	8.7	10.6	15.7	15.9	15.9	14.6	13.2	6.8	3.1
1969	6.1	1.7	4.2	8.4	12.4	14.7	18.2	17.3	14.3	13.3	6.1	3.6
1970	4.3	3.6	4.1	7.6	13.8	17.0	16.2	16.6	15.3	11.2	8.5	4.3
1971	4.9	4.7	5.5	8.2	12.4	13.1	18.2	16.5	14.7	11.6	6.4	6.7
1972	4.2	4.8	7.3	8.8	11.3	12.7	16.4	16.2	12.4	11.1	6.7	6.1
1973	4.5	4.5	6.6	8.2	12.2	15.9	16.5	17.7	15.4	9.2	6.5	5.3
1974	6.5	5.9	5.9	8.4	11.5	14.6	16.2	16.0	12.8	8.0	7.6	8.3
1975	7.2	4.9	5.2	9.1	10.4	15.5	18.6	19.6	14.3	10.4	6.2	4.6

1976	6.0	4.9	5.4	8.5	13.1	18.5	19.6	17.9	14.3	11.2	6.3	2.1
1977	3.1	6.1	7.4	7.7	10.9	12.6	16.5	16.0	14.0	11.9	6.7	6.4
1978	3.5	3.0	7.1	5.8	12.1	14.4	15.8	15.8	14.9	12.2	8.6	4.7
1979	0.1	1.5	5.3	8.6	11.0	14.9	17.8	16.0	14.1	11.8	7.2	6.4
1980	2.5	6.4	5.4	9.3	11.5	14.9	15.5	17.0	15.7	9.5	6.8	5.7
1981	4.9	3.2	8.8	8.2	12.0	14.3	16.8	17.5	15.4	9.0	7.9	0.8
1982	2.7	5.3	6.6	9.4	12.4	17.0	17.4	17.1	15.4	10.7	8.5	4.8
1983	7.3	2.1	6.8	7.8	11.3	15.3	21.1	18.4	14.6	11.0	7.8	6.0
1984	4.5	3.7	5.3	8.7	10.3	15.8	18.1	18.6	14.6	11.8	8.7	5.5
1985	1.0	2.4	5.2	9.1	11.5	13.3	17.2	15.6	15.3	11.2	4.6	7.3
1986	3.9	-1.6	5.7	6.7	11.9	15.6	17.2	14.7	12.1	11.6	8.4	6.5
1987	0.8	4.1	4.9	10.9	10.9	14.1	17.0	16.8	14.8	10.4	6.7	5.9
1988	5.7	5.2	7.0	8.6	12.5	14.5	15.3	16.2	14.2	11.3	5.4	7.8
1989	6.4	6.5	8.3	7.3	14.2	15.7	19.5	17.9	15.9	12.4	6.5	5.8
1990	7.3	8.3	8.9	8.5	13.5	14.5	18.0	19.4	14.1	12.6	7.1	4.3

Source: http://www.metoffice.gov.uk/climate/uk/stationdata/. (Contains public sector information licensed under the Open Government Licence).

Solution   To read data (which are also held within the file oxforddata1961-1990.csv) into R, use:

```
oxmetdata <- read.csv("wex11_1.csv")
```

This command creates an object from the data in the file; we have chosen to call the object oxmetdata. Using the command str to examine the object reveals that it is currently being stored as a dataframe, not a time series:

```
str(oxmetdata)
```

```
'data.frame':360 obs. of  4 variables:
$ yyyy : int   1961 1961 1961 1961 1961 1961
   1961 1961 1961 1961 …
$ mm    : int   1 2 3 4 5 6 7 8 9 10 …
$ tmean: num   4.2 7.6 8.7 10.7 11.8 15.6
   16.5 16.6 16 11.3 …
$ rain : num   86.1 53.4 4.5 75.9 28 33.9
   54.3 59.1 56.9 67.7 …
```

To create a time series object from the data contained in this dataframe, the ts command is used. This permits the specification of the start time and the frequency:

```
oxtemp <- ts(data = oxmetdata$tmean,
  start=c(1961,1), frequency=12)
```

Note that the frequency is given as the number of observations per year, so for monthly data frequency=12, for quarterly data frequency=4, and for annual data frequency=1, and so on. In the present case, the data begin in the first period of the year 1961 and so the start is specified as start=c(1961,1). For example, had the time series begun in June, we would have needed to specify start=c(1961,6) instead.

Use the is.ts() or str() command to confirm that the data are now held as a time series:

```
is.ts(oxtemp)
```

```
[1] TRUE
```

```
> str(oxtemp)
```

```
Time-Series [1:360] from 1961 to 1991: 4.2
   7.6 8.7 10.7 11.8 15.6 16.5 16.6 16 11.3 …
```

To plot the data, we can simply use the plot command. This command figures out that we are dealing with a time series and applies sensible defaults. The plot can be tidied up with appropriate axis labels as before:

```
plot(oxtemp, xlab="Year", ylab="Mean monthly
   temperature [°C]")
```

The result is as given in Figure 11.2.

11.2.3 Decomposing time series: trends, seasonality and irregular fluctuations

It is, of course, possible to calculate the mean, variance and standard deviation of a time series in exactly the same way as demonstrated in Chapter 2 for any other dataset. If the mean and variance of a time series are constant over the long-term,

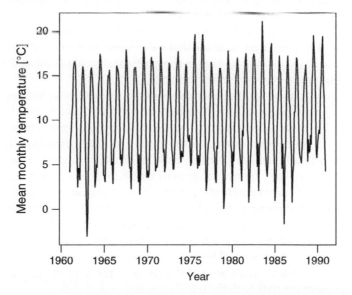

Figure 11.2 Time series of Oxford temperature data from Table 11.1 plotted as a time series using R. Source: Data from Met Office contains public sector information licensed under the Open Government Licence.

and if there are no periodic fluctuations, then the dataset can be described as *stationary*. Stationarity is a rare property in real-world datasets but the assumption of stationarity is fundamental to a number of advanced time series analysis techniques and many procedures are available for transforming non-stationary time series into stationary ones. Nevertheless, in geographical applications the non-stationary components of a time series are often of most interest. These features include trends, seasonality and other cyclical variations, which are now discussed in turn.

11.2.4 Analysing trends

The simplest patterns that can be observed in a time series are trends. We can say that there is a trend in our time series if there is a long-term change in the mean. How ought we to define long-term? There is no hard-and-fast rule, and a level of judgement is needed. Clearly in the CO_2 dataset presented above there is a long-term trend because the data continue going up and up even though there is also a seasonal cycle superimposed on the long-term pattern. However, were we to extend the dataset back in time (e.g. using measurements of CO_2 concentrations from ice cores) we would see that there have been cyclical variations in CO_2 concentration in the geological past as well as the strong trend in recent modern records.

Trends can be spotted relatively easily by using time series plots but it is important to be able to quantify and characterize the trend in order to understand its nature and origins. There are two ways to figure out the nature of the trend. The easiest is to hypothesize a functional form for the trend, such as a straight line or a parabola or an exponential growth or decay trend, which may accord with the theoretical expectation or be guessed at from the shape of the curve. Once a functional form has been identified, the fitted parameters of the function that described the trend can be found, together with error terms, using the techniques presented in Chapter 9. The residuals from such a regression represent the fluctuations around the trend in the dataset (i.e. they represent the data with the trend removed or detrended).

An alternative possibility is that the data display a trend but that it cannot be characterized using a simple functional form such as a straight line or an exponential. This situation may arise either because there is no simple function to describe the features present, or because a complex set of changes occurs over time in the dataset. In such cases it is often more useful to represent the long-term pattern using a moving average. A simple three-point moving average can be constructed by taking the average of the value of the dataset for the point in question together with the two points on either side (Figure 11.3). Adopting the same notation as in previous chapters, but letting x_t be an observation of quantity x at time t, the moving average can therefore be defined as:

$$y_t = \frac{1}{3}\sum_{r=-1}^{r=+1} x_{t+r}, \tag{11.1}$$

where r is an index used in the summation notation to indicate that we average the previous and next observations in the time series alongside the current observation, to find the moving average.

The measure defined in Equation 11.1 is the most basic kind of moving average. It looks both forward and backwards, and it treats each of the three component observations equally. In some circumstances more complex moving averages are required.

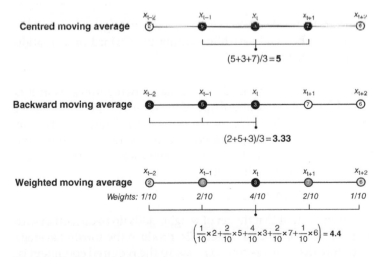

Figure 11.3 Types of moving average.

The simple moving average does not provide an estimate for the first and last time period in question. In situations where the subsequent observations are not available (e.g. in real-time monitoring systems) or in situations where only knowledge of the historical condition should contribute to the estimate of the moving average, a backward-looking moving average can be used. The backward-looking moving average bases the calculation only on observations that occurred prior to the current time period (Figure 11.3).

The standard moving average places equal emphasis on each of the three observations that contribute to the average for each data point. In some circumstances, it may be desirable to put extra emphasis on recent observations. This can be achieved by using a weighted moving average where observations that are closer in time to the average are given more emphasis or weight. Written generally, the formula for a weighted average is:

$$y_t = \sum_{r=-q}^{+s} a_t x_{t+r},\tag{11.2}$$

in which a_t is a set of weights that add up to one. The moving average represented by Equation 11.2 extends q units prior to the current observation and s units forward into the future.

Worked Example 11.2 Use a moving average to summarize possible trends and variability within the Oxford precipitation data presented in Table 11.2.

Solution The command to calculate moving averages in R is called filter. The size and shape of the filter are specified as a set of weights, which are identical to the set a_t in Equation 11.2. For a simple three-period, centred moving average we define the weights to count one-third equally from each of the three observations that we want to include:

```
a <- c(1/3,1/3,1/3)
```

It is important that the set of weights adds up to one, otherwise the average will not be correct. To produce the moving average and store the result as oxrain_ma3c the required command is:

```
oxrain_ma3c <- filter(oxrain, a, sides=2)
```

Table 11.2 Mean monthly precipitation data (mm/month) for Oxford, UK, 1961–1990.

	Jan	Feb	Mar	Apr	May	Jun	Jul	Aug	Sep	Oct	Nov	Dec
1961	86.1	53.4	4.5	75.9	28.0	33.9	54.3	59.1	56.9	67.7	36.9	97.8
1962	94.7	10.3	32.4	50.7	38.6	5.5	54.6	96.1	115.0	29.6	55.1	50.4
1963	27.9	10.4	75.9	50.4	43.6	65.9	48.0	81.4	58.0	44.4	115.4	18.6
1964	13.0	22.2	92.6	59.5	44.4	49.3	21.2	14.6	15.5	17.5	25.0	43.7
1965	56.0	5.8	44.2	49.4	63.8	53.4	90.0	61.7	75.4	12.0	51.2	95.5
1966	30.8	74.9	9.9	81.7	54.0	56.6	80.9	78.8	52.5	122.7	28.0	80.4
1967	35.5	64.3	30.7	40.4	85.6	45.3	76.6	44.0	59.4	121.7	29.2	63.3
1968	57.9	26.8	19.5	46.1	66.2	65.2	141.6	91.7	108.3	67.7	54.5	63.6
1969	74.9	39.9	48.3	37.0	88.8	20.7	49.0	87.9	32.8	5.0	52.0	72.9
1970	66.6	49.2	45.8	75.9	22.9	43.0	42.3	51.9	40.3	19.5	150.3	33.9
1971	105.3	22.1	42.9	44.8	46.3	135.8	37.3	133.2	13.4	74.2	61.3	24.0
1972	60.5	50.8	63.2	45.8	49.8	55.8	23.3	29.8	31.0	25.9	65.8	74.0
1973	24.4	15.0	10.8	42.4	71.2	109.3	60.0	28.8	32.7	40.8	28.3	30.9
1974	65.0	77.7	36.7	4.2	31.6	73.7	41.7	97.2	156.1	66.4	95.1	33.4

(Continued)

Table 11.2 (Continued)

	Jan	Feb	Mar	Apr	May	Jun	Jul	Aug	Sep	Oct	Nov	Dec
1975	69.4	29.7	95.2	46.9	49.0	18.0	34.1	30.2	95.3	11.0	37.9	21.7
1976	18.9	18.8	16.0	8.1	44.2	17.1	14.3	23.9	90.7	99.5	55.9	101.8
1977	75.9	114.4	59.8	40.7	43.1	86.9	5.8	129.0	7.8	32.8	46.6	67.8
1978	68.2	45.6	52.3	45.2	48.5	39.9	82.4	32.6	24.4	5.4	22.8	109.5
1979	56.8	56.2	110.3	38.9	124.8	43.6	16.6	60.8	17.0	59.8	39.4	127.7
1980	36.7	39.8	69.8	14.1	35.5	89.4	59.5	63.6	65.6	84.9	42.5	32.9
1981	39.2	20.9	129.7	43.9	87.7	26.1	34.0	38.8	92.5	64.7	31.4	74.0
1982	57.2	32.0	96.2	27.4	35.9	69.4	23.5	37.9	51.9	83.4	77.3	62.4
1983	45.7	23.8	42.5	89.5	106.3	19.4	21.7	24.0	59.4	53.0	42.6	49.8
1984	81.5	35.4	50.3	1.6	72.0	30.0	13.6	32.7	91.1	54.4	101.6	44.5
1985	44.4	34.7	36.7	32.7	80.5	122.5	45.8	67.0	19.4	29.1	41.1	108.1
1986	67.3	7.3	58.0	70.4	70.7	22.3	37.1	114.3	39.0	75.1	78.4	64.2
1987	9.9	34.0	53.9	51.4	48.7	86.0	29.3	34.0	25.4	138.8	58.2	26.0
1988	100.0	33.8	59.1	30.2	32.0	59.6	96.5	38.8	39.6	38.6	29.5	13.4
1989	33.4	55.1	47.0	63.8	29.2	45.9	29.4	43.3	23.8	53.9	32.5	141.5
1990	75.0	94.2	18.4	20.4	8.7	48.8	17.7	26.5	41.1	45.9	20.0	56.3

Source: http://www.metoffice.gov.uk/climate/uk/stationdata/. Contains public sector information licensed under the Open Government Licence.

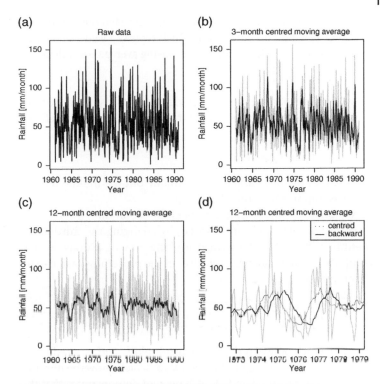

Figure 11.4 Moving averages of Oxford rainfall data: (a) raw data;
(b) three-point centred moving average with raw data shown in grey;
(c) 12-point centred moving average with raw data shown in grey; and
(d) comparison of centred (black dots) and backward (solid black line)
12-point moving averages with raw data shown in grey.

The first argument gives the dataset that we want to filter, the
second uses the filter specified above, and the final argument
indicates that we want a centred moving average. The average
can be plotted on top of the original time series with:

```
plot(oxrain, xlab="Year", ylab="Rainfall
    [mm/month]", col="grey", main="3-month
    centred moving average")

lines(oxrain_ma3c, xlab="Year",
    ylab="Rainfall [mm/month]")
```

The filter may be adjusted in several ways: first, its size can be
changed. A 12-point moving average is shown in Figure 11.4c. For

long-period moving averages, the weights can be tedious to type and so R provides the rep command, which repeats a value a specified number of times, making a 12-point moving average equivalent to:

```
a <- rep(1/12,12).
```

An example of a weighted moving average, which gives extra emphasis to observations closer to its centre can be defined thus:

```
a <- c(1/4,1/2,1/4)
```

Note that the filter still adds up to one but the middle term is twice as important as the other two.

Asymmetric filters can be used in situations where only historical observations are wanted, by changing the third argument to sides=1. A backward-looking moving average, which computes the average precipitation in the preceding year, can be obtained using:

```
a <- rep(1/12,12)
oxrain_ma12b <- filter(oxrain, a, sides=1)
```

The difference between the backward-looking and centred moving averages can be seen in Figure 11.4d, which shows a subset of the data between 1973 and 1979. This plot demonstrates that the centred moving average remains aligned with the original dataset but that, by definition, it cannot be computed at the ends of the time series. By contrast, the backward-looking moving average can be computed up until the very end of the time series (although not at the beginning) and it appears offset from the raw data, owing to its backward-looking nature (Figure 11.4d). In practice, the centred moving average is of more use to a climatologist wishing to relate precipitation variability to driving atmospheric processes; the backward-looking moving average is more useful to a water company official looking to see if a drought is on its way.

11.2.5 Removing trends ('detrending' data)

It is often necessary to remove trends from data, especially if the trend is not the feature of primary interest. For example, it is known that agricultural yields have systematically improved since the introduction of nitrogen-based fertilizers in the 1950s.

This trend is extremely interesting but once detected its presence obscures some other equally important observations about wheat yields. There are two simple ways to remove trends from time series data in order to highlight the fluctuations that are superimposed upon the trend. The first is to subtract the moving average from the original dataset, thereby creating a dataset from which the long-term trend has been removed. The second method is known as differencing. It involves subtracting the previous observation from the current observation in the time series, so that the new time series, y_t, is defined as:

$$y_t = x_t - x_{t-1}. \tag{11.3}$$

11.2.6 Quantifying seasonal variation

Seasonal variability is important in many geographical datasets encountered in physical and human geography. Seasonality typically has an annual period although the response of the system under study may exhibit variations on other timescales. In many applications, the seasonal cycle is the primary feature of the dataset. In such cases, which include many applications in climatology and ecology, the most straightforward way to summarize the seasonal cycle is to present monthly averages as anomalies from the mean. Thus, the seasonal cycle contained within a monthly time series can be characterized by taking the average of all values measured in January, February, March, and so on, relative to their respective annual means.

By contrast, there are some occasions when the seasonal cycle obscures more interesting features of the data, especially when the seasonality of a phenomenon is already known and well understood. A commonly cited example is in retail sales where seasonal variation is expected and it is usual to compare seasonally adjusted sales figures. Once the seasonal cycle has been identified it can be eliminated with a monthly differencing technique similar to that described in Section 11.2.4. The major difference is that the choice of differencing period must be made to match the period of the seasonal cycle. For example, if an annual cycle is discovered then the 12-month difference should be used, such that with monthly data:

$$y_t = x_t - x_{t-12}, \tag{11.4}$$

More sophisticated methods for seasonal adjustment are available to account for calendar effects and other peculiarities.

Once trends and seasonal variability have been diagnosed and analysed the remaining fluctuations in the dataset are left behind. These are of interest in the same way that residuals from a regression analysis are of interest, because they help to reveal notable anomalies from the principal modes of temporal variability. Fortunately, R is equipped with the ability to automate the decomposition of time series in to these three components: (i) trend; (ii) seasonality; and (ii) residual variability.

Worked Example 11.3 Calculate monthly temperature and rainfall climatologies for Oxford using the data in Table 11.1 and Table 11.2.

Solution The data are already in monthly increments, but to calculate a monthly climatology it is necessary to take an average across each of the months in turn (i.e. to produce an average across all Januaries, Februaries, etc.). The function to compute such a summary in R is called `tapply`, which applies a function to groups of entries in a table. Here we want to calculate the mean of groups that correspond to the months of the year, and so we create a new dataset called `oxtemp_month` by using tapply to calculate mean values according to where a particular observation is in the annual cycle. The command is:

```
oxtemp_month <- tapply(oxtemp,
  cycle(oxtemp), mean)
```

which returns a set of results (to two decimal places):

```
   1     2     3     4     5
 4.10  4.19  6.27  8.55 11.92

   6     7     8     9    10    11    12
15.03 17.11 16.74 14.43 11.10  6.90  4.94
```

The monthly temperatures can be plotted as a bar chart using the command:

```
barplot(oxtemp_month, names.arg=month.abb,
  xlab="Month", ylab="Temperature [°C]")
```

(a)

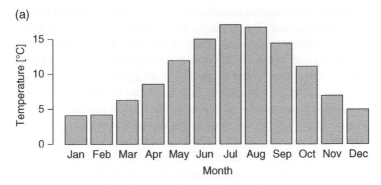

(b)

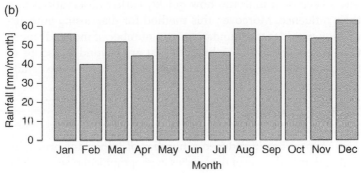

Figure 11.5 Monthly climatology for Oxford for (a) temperature and (b) precipitation.

which makes use of R's built-in list of abbreviated month names, month.abb. The resulting plot is shown in Figure 11.5aa along with its counterpart for rainfall (Figure 11.5b), which can be produced in a similar way.

11.2.7 Autocorrelation

A further noteworthy feature that may be present in time series data is autocorrelation. It is useful to know whether an observation made at time t is likely to be similar to one measured at time $t-1$. Autocorrelation commonly arises in many geographical situations. For example, in climatology observations of daily temperature often show a high level of autocorrelation, which suggests that tomorrow's weather will likely be similar to today's.

In order to quantify autocorrelation in a time series, autocorrelation coefficients can be calculated. These are equivalent to the correlation coefficient calculated in Chapter 8 but in the case of autocorrelation, observations are correlated with the values taken at different distances apart in the time series. In general, the correlation between values of the time series at k units apart is called the autocorrelation coefficient at lag k, and is written r_k. A plot showing correlation coefficients at lags of different lengths is called a *correlogram* (e.g. Figure 11.6). Care is necessary in the interpretation of the correlogram in the presence of trends and seasonality and in practice the correlogram is most usefully applied to datasets that have had trend and seasonality removed. The correlogram is a useful diagnostic of persistence in a time series because it indicates how quickly earlier observations lose their influence. Moreover, this method for diagnosing autocorrelation provides the foundation for a number of more advanced time series forecasting techniques that are beyond the scope of this book (see Diggle, 1990, and Chatfield, 1996, for more detail).

11.3   Summary

This chapter has introduced a number of important principles that underpin the use of time series in geographical research. The identification of trends, seasonality and examination of residual variability are all of primary importance in many applications in physical and human geography. These elementary techniques also pave the way for more advanced analyses which can be found in the many books on time series analysis. Advanced texts that are highly recommended for further reading include *Time Series: A Biostatistical Introduction* (Diggle, 1990) and *The Analysis of Time Series: An Introduction* (Chatfield, 1996). Both give good theoretical discussion alongside practical advice on how to undertake many of the analyses described in this chapter.

Exercises

1   Plot the corn yield data from Table 8.1 as a time series. Use a moving average to characterize the long-term variability in the dataset. Experiment with a range of averaging periods and comment on the impact that your choice of averaging period has on the effectiveness of the moving average as a summary of the dataset.

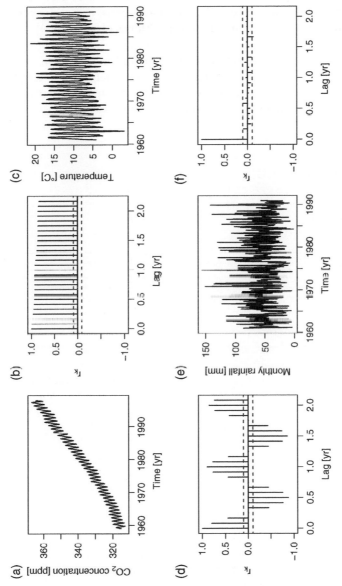

Figure 11.6 Typical datasets with their corresponding correlograms shown below: (a, b) CO_2 concentrations showing the effect of a trend on the correlogram; (c, d) Oxford temperature showing the effect of a seasonal pattern in the dataset; and (e, f) Oxford rainfall showing a dataset which naturally has very little autocorrelation. The dashed line indicates the threshold for statistically significant autocorrelation.

2 Monthly rainfall data in Oxford, UK are given in Table 11.2. Read the data into R and produce a time series plot using appropriate axis labels. Construct a seasonal climatology for the dataset and comment on the patterns observed.

3 Read the documentation on R's built-in time series analysis function `stl` which decomposes time series into trend, seasonality and remaining variability. Apply the procedure to the CO_2 dataset and comment on the results.

Appendix A: Introduction to the R package

A.1   Obtaining R

R is open source software, which is free to download and use. It is available from www.r-project.org. The R software is similar to S-PLUS, which is a commercial product. It has very wide support from professional and academic statisticians, and is well supported and documented both online and via textbooks (e.g. Venables and Ripley, 2003; Crawley, 2005, 2012).

The R software operates through a command-line interface: commands are typed into a window and the results are returned to the user through the same window, or are plotted if the commands involve graphics. Whilst this interface lacks some of the graphical appeal familiar to most modern computer users, it has the very important advantage that the series of commands that the user has issued can be saved for later reference and re-use, and that routine operations can be automated by placing commands in a text file and running them in one go.

A.2   Simple calculations

Simple calculations can be performed in the R command window. For example, to add two numbers, the user simply types:

```
>1 + 2
3
```

Statistical Analysis of Geographical Data: An Introduction, First Edition.
Simon J. Dadson.
© 2017 John Wiley & Sons Ltd. Published 2017 by John Wiley & Sons Ltd.

Here, the '>' sign is the R prompt, which indicates that R is waiting for the user to type something. The result '[1] 3' needs no explanation, except to note that the '[1]' which precedes the answer indicates that the number which follows is the first in the list of numbers in the answer. In this case it is the only number in the answer.

R can perform all of the basic arithmetic operations (and more besides), using '*' and '/' to represent multiplication and division, respectively, for example:

```
>10 – 7
[1]  3
>3*4
[1]  12
>25/5
[1]  5
```

Powers can be calculated using the '^' or '**' operators, and fractional powers are possible too, although a specific sqrt() function exists to calculate square roots:

```
>2^2
[1]  4
>5^2
[1]  25
>2**3
[1]  8
> sqrt(36)
[1]  6
>36^0.5
[1]  6
>36**0.5
[1]  6
>36**1.5
[1]  216
```

The results of any of these calculations can be stored in named variables using the assignment operator '<–' and recalled simply by reference to the variable name. For example:

```
> a<- 2+3
> b<- 10
> a
[1] 5
> b
[1] 10
```

Variables can then be used in subsequent calculations, so that:

```
> a/b
[1] 0.5
```

R refers to variables as 'objects', and a list of all currently stored objects can be generated with the ls() function:

```
> ls()
[1] "a" "b"
```

which, in this case, informs us of the two objects stored to date.

Variables can hold numerical data, but they can also hold characters, or logical (Boolean) values that are constrained to take the values TRUE or FALSE, abbreviated T or F. The class() function can be used to figure out which type of data is held in a particular variable. For example:

```
> class(a)
[1] "numeric"
> d<- T
> class(d)
[1] "logical"
> c<-"cat"
> c
[1] "cat"
> class(c)
[1] "character"
```

It is possible to return the length of an object using the length() function:

```
> length(a)
[1] 1
```

and any object can be removed from memory using the `rm()` function.

A.3 Vectors

In some cases, it is desirable to store a list of numbers. We could, of course, store a series of individual numbers in variables named a, b, c, and so on but when the numbers are themselves related it is easier to store them in a list called a vector. Vectors can be created in many ways in R. The simplest way is with the `c()` command, which stands for combine. Any number of values can be combined in this way:

```
> myvector<- c(1,2,3,4)
> myvector
[1] 1 2 3 4
```

Individual elements of a vector can be referred to using index notation, where the required index term is supplied in brackets after the name of the vector. So the third element of a vector is referred to as follows:

```
>myvector[3]
[1] 3
```

Standard vector arithmetic can be performed, including multiplication by and addition of a scalar:

```
> myvector * 2
[1] 2 4 6 8
> myvector+1
[1] 2 3 4 5
```

Quite frequently it is useful to be able to produce regular sequences of numbers. To do this, use the `seq()` command, which takes the arguments 'from' and 'to', so that `seq(1,10)` returns a vector containing the set of integers from 1 to 10.

```
> seq(1, 10)
[1] 1 2 3 4 5 6 7 8 9 10
```

To pick only even integers, for example, the optional argument 'by' can be employed:

```
> seq(2, 10, by=2)
[1]  2  4  6  8  10
```

A.4   Basic statistics

Once the concept of a vector is understood, statistical functions that operate on vectors are straightforward. Functions such as sum() and mean() can be used to perform their respective operations on numeric vectors:

```
> sum(myvector)
[1]  10
> mean(myvector)
[1]  2.5
> median(myvector)
[1]  2.5
```

More advanced statistical functions such as the variance and standard deviation are described as they are encountered in the main text.

A.5   Plotting data

Plotting is straightforward in R. Very fine control can be exerted over the output style and format. This facility takes a while to learn but offers a significant improvement over spreadsheet-style statistical packages especially when producing figures for reports and publications. A simple plot can be produced using the plot() command (Figure A.1). Note that plot() is part of R's graphics package, which must be loaded beforehand if it is not already using the command require(graphics).

```
> a<- c(1,2,3,4,5)
> b<- c(2,4,6,8,10)
```

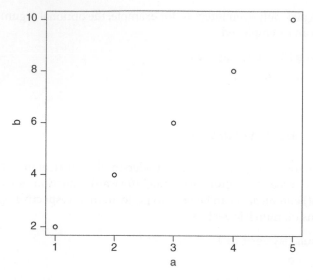

Figure A.1 A basic plot.

```
> require(graphics)
> plot(a,b)
```

Plots can be tailored by specifying additional options on the command line. Some of the more commonly used options are given in Table A.1.

An enormous number of other plotting options is documented in the R help [type help(plot) to see basic documentation and help(par) for details of advanced options]. An example which combines several of the options in Table A.1 is:

```
plot(a,b, type="b", xlab="Independent
variable", ylab="Dependent variable",
xlim=c(0,10), ylim=c(0,10), cex=1,
col="blue", pch=4, lty=2)
```

This command tells R to plot a and b, with type=b (i.e. both lines and points). The *x*- and *y*-axis labels are specified and the axis limits are chosen manually. The option cex=1 sets the size of the points which will be plotted, col="blue" sets the colour to blue (many other colours can be chosen conveniently by name, although a full range of colours is available to specify

Table A.1 Plotting options in R.

type	p = points l = lines b = both
xlab	Text for *x*-axis label
ylab	Text for *y*-axis label
pch	Symbol for plotting (see Figure A.2)
lty	Line type 1 = solid 2 = dashed 3 = dotted 4 = dot-dashed 5 = long dash 6 = long dot-dash
cex	symbol size (1 = default)
log	Log = "x" for log *x*-axis Log = "y" for log *y*-axis Log = "xy" for log–log plot
xlim, ylim	Sets range of *x*- and *y*-axis limits, for example xlim = c(0,10), ylim = c(0,10)

using hexadecimal notation). The option pch controls the specific symbol that is used to plot points (crosses in the present case; but see Figure A.2 for the full range of options), and the line style is controlled by `lty`. The finished result is presented in Figure A.3.

A.6 Multiple figures

Once a plot has been started with the `plot()` function, additional datasets may be added to the same figure by using the commands `points()` and `lines()` which take many of the same arguments as the main plot function. If this option is chosen, a legend can be added with the use of the `legend()` function which is illustrated in the example below.

Multiple plots can be created on the same page using the `mfrow` argument to the `par()` function. For example, `par(mfrow=c(2,2))` sets the page to receive a 2×2 set of plot (Figure A.4). Once this function has been called, the first plot is

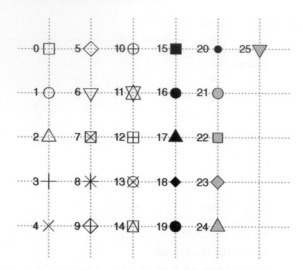

Figure A.2 List of plotting symbols commonly used in R graphics.

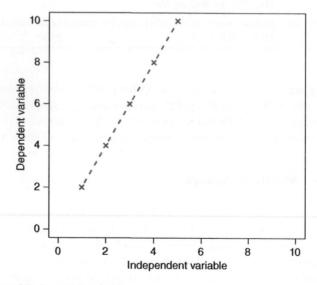

Figure A.3 A customized plot.

created in the top left-hand position. Subsequent plots occupy successive positions moving left-to-right and top-to-bottom. The following example illustrates several of the aforesaid features together, and produces output in publication-quality '.eps' format.

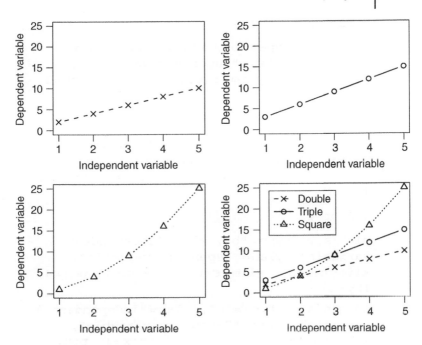

Figure A.4 Multiple plots on a page.

```
require(graphics)
postscript(file = "myfile.eps", one-
file = FALSE, horizontal = FALSE, width = 6,
height = 6)
par(mfrow = c(2,2))
a <- c(1,2,3,4,5)
b <- c(2,4,6,8,10)
c <- c(3,6,9,12,15)
d <- c(1,4,9,16,25)
# Plot 1
plot(a,b, type = "b", xlab = "Independent
variable", ylab = "Dependent variable",
xlim = c(1,5), ylim = c(0,25), cex = 1, pch = 4,
lty = 2)
```

```
# Plot 2

plot(a,c, type="b", xlab="Independent
variable", ylab="Dependent variable",
xlim=c(1,5), ylim=c(0,25), cex=1, pch=1,
lty=1)

# Plot 3

plot(a,d, type="b", xlab="Independent
variable", ylab="Dependent variable",
xlim=c(1,5), ylim=c(0,25), cex=1, pch=2,
lty=3)

# Plot 4

plot(a,b, type="b", xlab="Independent
variable", ylab="Dependent variable",
xlim=c(1,5), ylim=c(0,25), cex=1, pch=4,
lty=2)

points(a,c, type="b", xlab="Independent
variable", ylab="Dependent variable",
xlim=c(1,5), ylim=c(0,25), cex=1, pch=1,
lty=1)

points(a,d, type="b", xlab="Independent
variable", ylab="Dependent variable",
xlim=c(1,5), ylim=c(0,25), cex=1, pch=2,
lty=3)

#Legend

legend("topleft", c("Double","Triple",
"Square"), lty=c(2,1,3), pch=c(4,1,2))

dev.off()
```

R is endowed with a great variety of useful plotting routines, some of the most popular at the time of writing are contained within the ggplot2 package.

A.7   Reading and writing data

Many data formats can be read into R. The most straightforward are data contained within a text file. Such files are typically organized as space- or tab-delimited files (i.e. the columns are separated with spaces or tabs) and can be read into R with the command read.table(). A special case which occurs frequently in practice is that of the comma-separated variable file (often called a '.csv' file), in which columns of data are separated by commas. Such files can routinely be generated by spreadsheet software such as Microsoft Excel, amongst other packages. The ubiquity of files in the comma-separated variable format has led the developers of R to create a set of commands to deal with these files more efficiently. The command read.csv() can be used with minimal extra configuration to tell R to read in a text file which has been formatted with comma-separated variables.

For example, to read the crop yield data from Worked Example 7.1, the following command is given:

```
cropyield<- read.csv("cropyield.csv")
```

Many additional options can be specified – a full list is available by typing ?read.table. The most useful are 'header', which is a logical value indicating whether the first row contains variable names. Note that if the '.csv' file does not contain a header row, the variable names can be specified using the argument 'col.names'. The argument 'na.strings' can be used to specify the string(s) in the input data which will be interpreted as missing values, and 'nrows' and 'skip' control the number of rows to skip before starting to read data, and the number of rows of data to read in altogether, respectively.

The resulting dataframe, cropyield, contains two variables:

```
> names(cropyield)
[1] "Treatment" "Yield"
```

Note that a subset of the dataframe can be selected using subscript notation, so the first 10 elements of cropyield are:

```
> cropyield[1:10,]
Treatment Yield
1 A 9
2 A 11
3 A 10
4 A 9
5 A 6
```

```
6  A  10
7  A  10
8  A  5
9  A  9
10 A  11
```

The data can be aggregated using the aggregate() function, which applies a given function to subsets of the data grouped by factor, so that the means of crop yield by treatment are obtained using:

```
> aggregate(Yield, list(Treatment), mean)
Group.1 x
1  A  9.0
2  B  14.2
3  C  10.1
```

A.8 Summary

The introduction offered in this appendix is intended to demystify the most basic set of R concepts and commands so that the technical business of using the software does not interfere with the process of learning about statistics. The reader who is interested in becoming more proficient in the R language is recommended to look at the online documentation first, and then to consult one of the more advanced texts (e.g. Venables and Ripley, 2003; Crawley, 2005, 2012; R Core Team, 2015).

Appendix B: Statistical tables

Statistical Analysis of Geographical Data: An Introduction, First Edition.
Simon J. Dadson.
© 2017 John Wiley & Sons Ltd. Published 2017 by John Wiley & Sons Ltd.

Table B1 z-Table.

z	0.00	0.01	0.02	0.03	0.04	0.05	0.06	0.07	0.08	0.09
0.0	0.5000	0.5040	0.5080	0.5120	0.5160	0.5199	0.5239	0.5279	0.5319	0.5359
0.1	0.5398	0.5438	0.5478	0.5517	0.5557	0.5596	0.5636	0.5675	0.5714	0.5753
0.2	0.5793	0.5832	0.5871	0.5910	0.5948	0.5987	0.6026	0.6064	0.6103	0.6141
0.3	0.6179	0.6217	0.6255	0.6293	0.6331	0.6368	0.6406	0.6443	0.6480	0.6517
0.4	0.6554	0.6591	0.6628	0.6664	0.6700	0.6736	0.6772	0.6808	0.6844	0.6879
0.5	0.6915	0.6950	0.6985	0.7019	0.7054	0.7088	0.7123	0.7157	0.7190	0.7224
0.6	0.7257	0.7291	0.7324	0.7357	0.7389	0.7422	0.7454	0.7486	0.7517	0.7549
0.7	0.7580	0.7611	0.7642	0.7673	0.7704	0.7734	0.7764	0.7794	0.7823	0.7852
0.8	0.7881	0.7910	0.7939	0.7967	0.7995	0.8023	0.8051	0.8078	0.8106	0.8133
0.9	0.8159	0.8186	0.8212	0.8238	0.8264	0.8289	0.8315	0.8340	0.8365	0.8389
1.0	0.8413	0.8438	0.8461	0.8485	0.8508	0.8531	0.8554	0.8577	0.8599	0.8621
1.1	0.8643	0.8665	0.8686	0.8708	0.8729	0.8749	0.8770	0.8790	0.8810	0.8830
1.2	0.8849	0.8869	0.8888	0.8907	0.8925	0.8944	0.8962	0.8980	0.8997	0.9015

(*Continued*)

Table B.1 (Continued)

z	0.00	0.01	0.02	0.03	0.04	0.05	0.06	0.07	0.08	0.09
1.3	0.9032	0.9049	0.9066	0.9082	0.9099	0.9115	0.9131	0.9147	0.9162	0.9177
1.4	0.9192	0.9207	0.9222	0.9236	0.9251	0.9265	0.9279	0.9292	0.9306	0.9319
1.5	0.9332	0.9345	0.9357	0.9370	0.9382	0.9394	0.9406	0.9418	0.9429	0.9441
1.6	0.9452	0.9463	0.9474	0.9484	0.9495	0.9505	0.9515	0.9525	0.9535	0.9545
1.7	0.9554	0.9564	0.9573	0.9582	0.9591	0.9599	0.9608	0.9616	0.9625	0.9633
1.8	0.9641	0.9649	0.9656	0.9664	0.9671	0.9678	0.9686	0.9693	0.9699	0.9706
1.9	0.9713	0.9719	0.9726	0.9732	0.9738	0.9744	0.9750	0.9756	0.9761	0.9767
2.0	0.9772	0.9778	0.9783	0.9788	0.9793	0.9798	0.9803	0.9808	0.9812	0.9817
2.1	0.9821	0.9826	0.9830	0.9834	0.9838	0.9842	0.9846	0.9850	0.9854	0.9857
2.2	0.9861	0.9864	0.9868	0.9871	0.9875	0.9878	0.9881	0.9884	0.9887	0.9890
2.3	0.9893	0.9896	0.9898	0.9901	0.9904	0.9906	0.9909	0.9911	0.9913	0.9916

2.4	0.9918	0.9920	0.9922	0.9925	0.9927	0.9929	0.9931	0.9932	0.9934	0.9936
2.5	0.9938	0.9940	0.9941	0.9943	0.9945	0.9946	0.9948	0.9949	0.9951	0.9952
2.6	0.9953	0.9955	0.9956	0.9957	0.9959	0.9960	0.9961	0.9962	0.9963	0.9964
2.7	0.9965	0.9966	0.9967	0.9968	0.9969	0.9970	0.9971	0.9972	0.9973	0.9974
2.8	0.9974	0.9975	0.9976	0.9977	0.9977	0.9978	0.9979	0.9979	0.9980	0.9981
2.9	0.9981	0.9982	0.9982	0.9983	0.9984	0.9984	0.9985	0.9985	0.9986	0.9986
3.0	0.9987	0.9987	0.9987	0.9988	0.9988	0.9989	0.9989	0.9989	0.9990	0.9990

To use the table, look up the first two digits of the z-score in the left-hand column, then read across to find the third digit.

For example, for 2.50 go down until you get to 2.5, the third digit of your z-score is 0 so the answer is 0.9938.

Again, for 1.54 go down until you get to 1.5, then across to 0.04 and the answer is 0.9382.

If the z-score is negative, remove the minus sign and use the table as normal but subtract your answer from one to get the final probability.

Table B.2 *t*-Distribution.

df, $\nu = n-1$	Significance, α				
	0.05	0.025	0.01	0.005	0.001
1	6.31	12.71	31.82	63.66	636.62
2	2.92	4.30	6.96	9.92	31.60
3	2.35	3.18	4.54	5.84	12.92
4	2.13	2.78	3.75	4.60	8.61
5	2.02	2.57	3.36	4.03	6.87
6	1.94	2.45	3.14	3.71	5.96
7	1.89	2.36	3.00	3.50	5.41
8	1.86	2.31	2.90	3.36	5.04
9	1.83	2.26	2.82	3.25	4.78
10	1.81	2.23	2.76	3.17	4.59
11	1.80	2.20	2.72	3.11	4.44
12	1.78	2.18	2.68	3.05	4.32
13	1.77	2.16	2.65	3.01	4.22
14	1.76	2.14	2.62	2.98	4.14
15	1.75	2.13	2.60	2.95	4.07
16	1.75	2.12	2.58	2.92	4.01
17	1.74	2.11	2.57	2.90	3.97
18	1.73	2.10	2.55	2.88	3.92
19	1.73	2.09	2.54	2.86	3.88
20	1.72	2.09	2.53	2.85	3.85
21	1.72	2.08	2.52	2.83	3.82
22	1.72	2.07	2.51	2.82	3.79
23	1.71	2.07	2.50	2.81	3.77
24	1.71	2.06	2.49	2.80	3.75
25	1.71	2.06	2.49	2.79	3.73
26	1.71	2.06	2.48	2.78	3.71
27	1.70	2.05	2.47	2.77	3.69
28	1.70	2.05	2.47	2.76	3.67

Table B.2 (Continued)

df, $\nu = n-1$	Significance, α				
	0.05	0.025	0.01	0.005	0.001
29	1.70	2.05	2.46	2.76	3.66
30	1.70	2.04	2.46	2.75	3.65
35	1.69	2.03	2.44	2.72	3.59
40	1.68	2.02	2.42	2.70	3.55
45	1.68	2.01	2.41	2.69	3.52
50	1.68	2.01	2.40	2.68	3.50
60	1.67	2.00	2.39	2.66	3.46
70	1.67	1.99	2.38	2.65	3.44
80	1.66	1.99	2.37	2.64	3.42
90	1.66	1.99	2.37	2.63	3.40
100	1.66	1.98	2.36	2.63	3.39
z	1.64	1.96	2.33	2.58	3.29

Remember to divide α by two for two-sided confidence intervals and two-tailed tests.

Formula for confidence interval around the mean:

$$\overline{x} - t_{\frac{\alpha}{2}, n-1} \frac{s}{\sqrt{n}} \leq \mu \leq \overline{x} + t_{\frac{\alpha}{2}, n-1} \frac{s}{\sqrt{n}}$$

Table B.3 Critical values of the F-distribution.

Critical values of the F distribution at 0.1 significance level

ν_2	Degrees of freedom, ν_1																		
	1	2	3	4	5	6	7	8	9	10	11	12	13	14	15	16	17	18	19
1	39.86	49.50	53.59	55.83	57.24	58.20	58.91	59.44	59.86	60.19	60.47	60.71	60.90	61.07	61.22	61.35	61.46	61.57	61.66
2	8.53	9.00	9.16	9.24	9.29	9.33	9.35	9.37	9.38	9.39	9.40	9.41	9.41	9.42	9.42	9.43	9.43	9.44	9.44
3	5.54	5.46	5.39	5.34	5.31	5.28	5.27	5.25	5.24	5.23	5.22	5.22	5.21	5.20	5.20	5.20	5.19	5.19	5.19
4	4.54	4.32	4.19	4.11	4.05	4.01	3.98	3.95	3.94	3.92	3.91	3.90	3.89	3.88	3.87	3.86	3.86	3.85	3.85
5	4.06	3.78	3.62	3.52	3.45	3.40	3.37	3.34	3.32	3.30	3.28	3.27	3.26	3.25	3.24	3.23	3.22	3.22	3.21
6	3.78	3.46	3.29	3.18	3.11	3.05	3.01	2.98	2.96	2.94	2.92	2.90	2.89	2.88	2.87	2.86	2.85	2.85	2.84
7	3.59	3.26	3.07	2.96	2.88	2.83	2.78	2.75	2.72	2.70	2.68	2.67	2.65	2.64	2.63	2.62	2.61	2.61	2.60
8	3.46	3.11	2.92	2.81	2.73	2.67	2.62	2.59	2.56	2.54	2.52	2.50	2.49	2.48	2.46	2.45	2.45	2.44	2.43
9	3.36	3.01	2.81	2.69	2.61	2.55	2.51	2.47	2.44	2.42	2.40	2.38	2.36	2.35	2.34	2.33	2.32	2.31	2.30
10	3.29	2.92	2.73	2.61	2.52	2.46	2.41	2.38	2.35	2.32	2.30	2.28	2.27	2.26	2.24	2.23	2.22	2.22	2.21
11	3.23	2.86	2.66	2.54	2.45	2.39	2.34	2.30	2.27	2.25	2.23	2.21	2.19	2.18	2.17	2.16	2.15	2.14	2.13
12	3.18	2.81	2.61	2.48	2.39	2.33	2.28	2.24	2.21	2.19	2.17	2.15	2.13	2.12	2.10	2.09	2.08	2.08	2.07
13	3.14	2.76	2.56	2.43	2.35	2.28	2.23	2.20	2.16	2.14	2.12	2.10	2.08	2.07	2.05	2.04	2.03	2.02	2.01
14	3.10	2.73	2.52	2.39	2.31	2.24	2.19	2.15	2.12	2.10	2.07	2.05	2.04	2.02	2.01	2.00	1.99	1.98	1.97
15	3.07	2.70	2.49	2.36	2.27	2.21	2.16	2.12	2.09	2.06	2.04	2.02	2.00	1.99	1.97	1.96	1.95	1.94	1.93
16	3.05	2.67	2.46	2.33	2.24	2.18	2.13	2.09	2.06	2.03	2.01	1.99	1.97	1.95	1.94	1.93	1.92	1.91	1.90
17	3.03	2.64	2.44	2.31	2.22	2.15	2.10	2.06	2.03	2.00	1.98	1.96	1.94	1.93	1.91	1.90	1.89	1.88	1.87

Degrees of freedom, ν_2

18	3.01	2.62	2.42	2.29	2.20	2.13	2.08	2.04	2.00	1.98	1.95	1.93	1.92	1.90	1.89	1.87	1.86	1.85	1.84
19	2.99	2.61	2.40	2.27	2.18	2.11	2.05	2.02	1.98	1.96	1.93	1.91	1.89	1.88	1.86	1.85	1.84	1.83	1.82
20	2.97	2.59	2.38	2.25	2.16	2.09	2.04	2.00	1.96	1.94	1.91	1.89	1.87	1.86	1.84	1.83	1.82	1.81	1.80
21	2.96	2.57	2.36	2.23	2.14	2.08	2.02	1.98	1.95	1.92	1.90	1.87	1.86	1.84	1.83	1.81	1.80	1.79	1.78
22	2.95	2.56	2.35	2.22	2.13	2.06	2.01	1.97	1.93	1.90	1.88	1.86	1.84	1.82	1.81	1.80	1.79	1.78	1.77
23	2.94	2.55	2.34	2.21	2.11	2.05	1.99	1.95	1.92	1.89	1.87	1.84	1.83	1.81	1.80	1.78	1.77	1.76	1.75
24	2.93	2.54	2.33	2.19	2.10	2.04	1.98	1.94	1.91	1.88	1.85	1.83	1.81	1.80	1.78	1.77	1.76	1.75	1.74
25	2.92	2.53	2.32	2.18	2.09	2.02	1.97	1.93	1.89	1.87	1.84	1.82	1.80	1.79	1.77	1.76	1.75	1.74	1.73
26	2.91	2.52	2.31	2.17	2.08	2.01	1.96	1.92	1.88	1.86	1.83	1.81	1.79	1.77	1.76	1.75	1.73	1.72	1.71
27	2.90	2.51	2.30	2.17	2.07	2.00	1.95	1.91	1.87	1.85	1.82	1.80	1.78	1.75	1.75	1.74	1.72	1.71	1.70
28	2.89	2.50	2.29	2.16	2.06	2.00	1.94	1.90	1.87	1.84	1.81	1.79	1.77	1.75	1.74	1.73	1.71	1.70	1.69
29	2.89	2.50	2.28	2.15	2.06	1.99	1.93	1.89	1.86	1.83	1.80	1.78	1.76	1.75	1.73	1.72	1.71	1.69	1.68
30	2.88	2.49	2.28	2.14	2.05	1.98	1.93	1.88	1.85	1.82	1.79	1.77	1.75	1.74	1.72	1.71	1.70	1.69	1.68
40	2.84	2.44	2.23	2.09	2.00	1.93	1.87	1.83	1.79	1.75	1.74	1.71	1.70	1.68	1.66	1.65	1.64	1.62	1.61
50	2.81	2.41	2.20	2.06	1.97	1.90	1.84	1.80	1.76	1.73	1.70	1.68	1.66	1.64	1.63	1.61	1.60	1.59	1.58
60	2.79	2.39	2.18	2.04	1.95	1.87	1.82	1.77	1.74	1.71	1.68	1.66	1.64	1.62	1.60	1.59	1.58	1.56	1.55
70	2.78	2.38	2.16	2.03	1.93	1.86	1.80	1.76	1.72	1.69	1.66	1.64	1.62	1.60	1.59	1.57	1.56	1.55	1.54
80	2.77	2.37	2.15	2.02	1.92	1.85	1.79	1.75	1.71	1.68	1.65	1.63	1.61	1.59	1.57	1.56	1.55	1.53	1.52
90	2.76	2.36	2.15	2.01	1.91	1.84	1.78	1.74	1.70	1.67	1.64	1.62	1.60	1.58	1.56	1.55	1.54	1.52	1.51
100	2.76	2.36	2.14	2.00	1.91	1.83	1.78	1.73	1.69	1.66	1.64	1.61	1.59	1.57	1.56	1.54	1.53	1.52	1.50
110	2.75	2.35	2.13	2.00	1.90	1.83	1.77	1.73	1.69	1.66	1.63	1.61	1.59	1.57	1.55	1.54	1.52	1.51	1.50
120	2.75	2.35	2.13	1.99	1.90	1.82	1.77	1.72	1.68	1.65	1.63	1.60	1.58	1.56	1.55	1.53	1.52	1.50	1.49
10000	2.71	2.30	2.08	1.95	1.85	1.77	1.72	1.67	1.63	1.60	1.57	1.55	1.52	1.51	1.49	1.47	1.46	1.44	1.43

Reject H_0 if the calculated value of F is greater than the critical value in the table.

Table B.3 (Continued)

Critical values of the F distribution at 0.1 significance level

ν_2	Degrees of freedom, ν_1																				
	20	21	22	23	24	25	26	27	28	29	30	40	50	60	70	80	90	100	110	120	10000
1	61.74	61.81	61.88	61.95	62.00	62.05	62.10	62.15	62.19	62.23	62.26	62.53	62.69	62.79	62.87	62.93	62.97	63.01	63.04	63.06	63.32
2	9.44	9.44	9.45	9.45	9.45	9.45	9.45	9.45	9.46	9.46	9.46	9.47	9.47	9.47	9.48	9.48	9.48	9.48	9.48	9.48	9.49
3	5.18	5.18	5.18	5.18	5.18	5.17	5.17	5.17	5.17	5.17	5.17	5.16	5.15	5.15	5.15	5.15	5.15	5.14	5.14	5.14	5.13
4	3.84	3.84	3.84	3.83	3.83	3.83	3.83	3.82	3.82	3.82	3.82	3.80	3.80	3.79	3.79	3.78	3.78	3.78	3.78	3.78	3.76
5	3.21	3.20	3.20	3.19	3.19	3.19	3.18	3.18	3.18	3.18	3.17	3.16	3.15	3.14	3.14	3.13	3.13	3.13	3.12	3.12	3.11
6	2.84	2.83	2.83	2.82	2.82	2.81	2.81	2.81	2.81	2.80	2.80	2.78	2.77	2.76	2.76	2.75	2.75	2.75	2.74	2.74	2.72
7	2.59	2.59	2.58	2.58	2.58	2.57	2.57	2.56	2.56	2.56	2.56	2.54	2.52	2.51	2.51	2.50	2.50	2.50	2.49	2.49	2.47
8	2.42	2.42	2.41	2.41	2.40	2.40	2.40	2.39	2.39	2.39	2.38	2.36	2.35	2.34	2.33	2.33	2.32	2.32	2.32	2.32	2.29
9	2.30	2.29	2.29	2.28	2.28	2.27	2.27	2.26	2.26	2.26	2.25	2.23	2.22	2.21	2.20	2.20	2.19	2.19	2.19	2.18	2.16
10	2.20	2.19	2.19	2.18	2.18	2.17	2.17	2.17	2.16	2.16	2.16	2.13	2.12	2.11	2.10	2.10	2.09	2.09	2.08	2.08	2.06
11	2.12	2.12	2.11	2.11	2.10	2.10	2.09	2.09	2.08	2.08	2.08	2.05	2.04	2.03	2.02	2.01	2.01	2.01	2.00	2.00	1.97
12	2.06	2.05	2.05	2.04	2.04	2.03	2.03	2.02	2.02	2.01	2.01	1.99	1.97	1.96	1.95	1.95	1.94	1.94	1.93	1.93	1.90
13	2.01	2.00	1.99	1.99	1.98	1.98	1.97	1.97	1.96	1.96	1.96	1.93	1.92	1.90	1.90	1.89	1.89	1.88	1.88	1.88	1.85
14	1.96	1.96	1.95	1.94	1.94	1.93	1.93	1.92	1.92	1.92	1.91	1.89	1.87	1.86	1.85	1.84	1.84	1.83	1.83	1.83	1.80
15	1.92	1.92	1.91	1.90	1.90	1.89	1.89	1.88	1.88	1.88	1.87	1.85	1.83	1.82	1.81	1.80	1.80	1.79	1.79	1.79	1.76
16	1.89	1.88	1.88	1.87	1.87	1.86	1.86	1.85	1.85	1.84	1.84	1.81	1.79	1.78	1.77	1.77	1.77	1.76	1.75	1.75	1.72
17	1.86	1.86	1.85	1.84	1.84	1.83	1.83	1.82	1.82	1.81	1.81	1.78	1.76	1.75	1.74	1.74	1.73	1.73	1.72	1.72	1.69

Degrees of freedom, ν_2

18	1.66	1.69	1.69	1.70	1.70	1.71	1.71	1.72	1.74	1.75	1.78	1.79	1.79	1.80	1.80	1.80	1.81	1.82	1.82	1.83	1.84
19	1.63	1.67	1.67	1.67	1.68	1.68	1.69	1.70	1.71	1.73	1.76	1.76	1.77	1.77	1.78	1.78	1.79	1.79	1.80	1.81	1.81
20	1.61	1.64	1.65	1.65	1.65	1.66	1.67	1.68	1.69	1.71	1.74	1.74	1.75	1.75	1.76	1.76	1.77	1.77	1.78	1.79	1.79
21	1.59	1.62	1.63	1.63	1.63	1.64	1.65	1.66	1.67	1.69	1.72	1.72	1.73	1.73	1.74	1.74	1.75	1.75	1.76	1.77	1.78
22	1.57	1.60	1.61	1.61	1.62	1.62	1.63	1.64	1.65	1.67	1.70	1.72	1.71	1.72	1.72	1.73	1.73	1.74	1.74	1.75	1.76
23	1.55	1.59	1.59	1.59	1.60	1.61	1.60	1.62	1.64	1.66	1.69	1.69	1.69	1.70	1.70	1.71	1.72	1.72	1.73	1.74	1.74
24	1.53	1.57	1.57	1.58	1.58	1.59	1.60	1.61	1.62	1.64	1.67	1.63	1.68	1.69	1.69	1.70	1.70	1.71	1.71	1.72	1.73
25	1.52	1.56	1.56	1.56	1.57	1.58	1.58	1.59	1.61	1.63	1.66	1.66	1.67	1.67	1.68	1.68	1.69	1.70	1.70	1.71	1.72
26	1.50	1.54	1.55	1.55	1.56	1.56	1.57	1.58	1.59	1.61	1.65	1.65	1.66	1.66	1.67	1.67	1.68	1.68	1.69	1.70	1.71
27	1.49	1.53	1.53	1.54	1.54	1.55	1.56	1.57	1.58	1.60	1.64	1.64	1.64	1.65	1.65	1.66	1.67	1.67	1.68	1.69	1.70
28	1.48	1.52	1.52	1.53	1.53	1.54	1.55	1.56	1.57	1.59	1.63	1.63	1.63	1.64	1.64	1.65	1.66	1.66	1.67	1.68	1.69
29	1.47	1.51	1.51	1.52	1.52	1.53	1.54	1.55	1.56	1.58	1.62	1.62	1.62	1.63	1.63	1.64	1.65	1.65	1.66	1.67	1.68
30	1.46	1.50	1.50	1.51	1.51	1.52	1.53	1.54	1.55	1.57	1.61	1.61	1.62	1.62	1.63	1.63	1.64	1.64	1.65	1.66	1.67
40	1.38	1.42	1.43	1.43	1.44	1.45	1.46	1.47	1.48	1.51	1.54	1.55	1.55	1.56	1.56	1.57	1.57	1.58	1.59	1.60	1.61
50	1.33	1.38	1.38	1.39	1.39	1.40	1.41	1.42	1.44	1.46	1.50	1.51	1.51	1.52	1.52	1.53	1.54	1.54	1.55	1.56	1.57
60	1.29	1.35	1.35	1.36	1.36	1.37	1.38	1.40	1.41	1.44	1.48	1.48	1.49	1.49	1.50	1.50	1.51	1.52	1.53	1.53	1.54
70	1.27	1.32	1.33	1.34	1.34	1.35	1.36	1.37	1.39	1.42	1.46	1.46	1.47	1.47	1.48	1.49	1.49	1.50	1.51	1.52	1.53
80	1.25	1.31	1.31	1.32	1.33	1.33	1.34	1.36	1.38	1.40	1.44	1.45	1.45	1.46	1.47	1.47	1.48	1.49	1.49	1.50	1.51
90	1.23	1.29	1.30	1.30	1.31	1.32	1.33	1.35	1.36	1.39	1.43	1.44	1.44	1.45	1.45	1.45	1.47	1.48	1.48	1.49	1.50
100	1.22	1.28	1.29	1.29	1.30	1.31	1.32	1.34	1.35	1.38	1.42	1.43	1.43	1.44	1.45	1.45	1.46	1.47	1.48	1.48	1.49
110	1.20	1.27	1.28	1.28	1.29	1.30	1.31	1.33	1.35	1.37	1.42	1.43	1.43	1.43	1.44	1.45	1.45	1.46	1.47	1.48	1.49
120	1.19	1.26	1.27	1.28	1.28	1.29	1.31	1.32	1.34	1.37	1.41	1.42	1.42	1.43	1.43	1.44	1.45	1.46	1.46	1.47	1.48
10000	1.00	1.17	1.18	1.19	1.20	1.21	1.22	1.24	1.26	1.30	1.34	1.35	1.35	1.36	1.37	1.38	1.38	1.39	1.40	1.41	1.42

Table B.3 (Continued)

Critical values of the F distribution at 0.05 significance level

Degrees of freedom, ν_1

ν_2	1	2	3	4	5	6	7	8	9	10	11	12	13	14	15	16	17	18	19
1	161.4	199.5	215.7	224.6	230.2	234.0	236.8	238.9	240.5	241.9	243.0	243.9	244.7	245.4	245.9	246.5	246.9	247.3	247.7
2	18.51	19.00	19.16	19.25	19.30	19.33	19.35	19.37	19.38	19.40	19.40	19.41	19.42	19.42	19.43	19.43	19.44	19.44	19.44
3	10.13	9.55	9.28	9.12	9.01	8.94	8.89	8.85	8.81	8.79	8.76	8.74	8.73	8.71	8.70	8.69	8.68	8.67	8.67
4	7.71	6.94	6.59	6.39	6.26	6.16	6.09	6.04	6.00	5.96	5.94	5.91	5.89	5.87	5.86	5.84	5.83	5.82	5.81
5	6.61	5.79	5.41	5.19	5.05	4.95	4.88	4.82	4.77	4.74	4.70	4.68	4.66	4.64	4.62	4.60	4.59	4.58	4.57
6	5.99	5.14	4.76	4.53	4.39	4.28	4.21	4.15	4.10	4.06	4.03	4.00	3.98	3.96	3.94	3.92	3.91	3.90	3.88
7	5.59	4.74	4.35	4.12	3.97	3.87	3.79	3.73	3.68	3.64	3.60	3.57	3.55	3.53	3.51	3.49	3.48	3.47	3.46
8	5.32	4.46	4.07	3.84	3.69	3.58	3.50	3.44	3.39	3.35	3.31	3.28	3.26	3.24	3.22	3.20	3.19	3.17	3.16
9	5.12	4.26	3.86	3.63	3.48	3.37	3.29	3.23	3.18	3.14	3.10	3.07	3.05	3.03	3.01	2.99	2.97	2.96	2.95
10	4.96	4.10	3.71	3.48	3.33	3.22	3.14	3.07	3.02	2.98	2.94	2.91	2.89	2.86	2.85	2.83	2.81	2.80	2.79
11	4.84	3.98	3.59	3.36	3.20	3.09	3.01	2.95	2.90	2.85	2.82	2.79	2.76	2.74	2.72	2.70	2.69	2.67	2.66
12	4.75	3.89	3.49	3.26	3.11	3.00	2.91	2.85	2.80	2.75	2.72	2.69	2.66	2.64	2.62	2.60	2.58	2.57	2.56
13	4.67	3.81	3.41	3.18	3.03	2.92	2.83	2.77	2.71	2.67	2.63	2.60	2.58	2.55	2.53	2.51	2.50	2.48	2.47
14	4.60	3.74	3.34	3.11	2.96	2.85	2.76	2.70	2.65	2.60	2.57	2.53	2.51	2.48	2.46	2.44	2.43	2.41	2.40
15	4.54	3.68	3.29	3.06	2.90	2.79	2.71	2.64	2.59	2.54	2.51	2.48	2.45	2.42	2.40	2.38	2.37	2.35	2.34
16	4.49	3.63	3.24	3.01	2.85	2.74	2.66	2.59	2.54	2.49	2.46	2.42	2.40	2.37	2.35	2.33	2.32	2.30	2.29
17	4.45	3.59	3.20	2.96	2.81	2.70	2.61	2.55	2.49	2.45	2.41	2.38	2.35	2.33	2.31	2.29	2.27	2.26	2.24

Degrees of freedom, ν_2

18	4.41	3.55	3.16	2.93	2.77	2.66	2.58	2.51	2.46	2.41	2.37	2.34	2.31	2.29	2.27	2.25	2.23	2.22	2.20
19	4.38	3.52	3.13	2.90	2.74	2.63	2.54	2.48	2.42	2.38	2.34	2.31	2.28	2.26	2.23	2.21	2.20	2.18	2.17
20	4.35	3.49	3.10	2.87	2.71	2.60	2.51	2.45	2.39	2.35	2.31	2.28	2.25	2.22	2.20	2.18	2.17	2.15	2.14
21	4.32	3.47	3.07	2.84	2.68	2.57	2.49	2.42	2.37	2.32	2.28	2.25	2.22	2.20	2.18	2.16	2.14	2.12	2.11
22	4.30	3.44	3.05	2.82	2.66	2.55	2.46	2.40	2.34	2.30	2.26	2.23	2.20	2.17	2.15	2.13	2.11	2.10	2.08
23	4.28	3.42	3.03	2.80	2.64	2.53	2.44	2.37	2.32	2.27	2.24	2.20	2.18	2.15	2.13	2.11	2.09	2.08	2.06
24	4.26	3.40	3.01	2.78	2.62	2.51	2.42	2.36	2.30	2.25	2.22	2.18	2.15	2.13	2.11	2.09	2.07	2.05	2.04
25	4.24	3.39	2.99	2.76	2.60	2.49	2.40	2.34	2.28	2.24	2.20	2.16	2.14	2.11	2.09	2.07	2.05	2.04	2.02
26	4.23	3.37	2.98	2.74	2.59	2.47	2.39	2.32	2.27	2.22	2.18	2.15	2.12	2.09	2.07	2.05	2.03	2.02	2.00
27	4.21	3.35	2.96	2.73	2.57	2.46	2.37	2.31	2.25	2.20	2.17	2.13	2.10	2.08	2.06	2.04	2.02	2.00	1.99
28	4.20	3.34	2.95	2.71	2.56	2.45	2.36	2.29	2.24	2.19	2.15	2.12	2.09	2.06	2.04	2.02	2.00	1.99	1.97
29	4.18	3.33	2.93	2.70	2.55	2.43	2.35	2.28	2.22	2.18	2.14	2.10	2.08	2.05	2.03	2.01	1.99	1.97	1.96
30	4.17	3.32	2.92	2.69	2.53	2.42	2.33	2.27	2.21	2.16	2.13	2.09	2.06	2.04	2.01	1.99	1.98	1.96	1.95
40	4.08	3.23	2.84	2.61	2.45	2.34	2.25	2.18	2.12	2.08	2.04	2.00	1.97	1.95	1.92	1.90	1.89	1.87	1.85
50	4.03	3.18	2.79	2.56	2.40	2.29	2.20	2.13	2.07	2.03	1.99	1.95	1.92	1.89	1.87	1.85	1.83	1.81	1.80
60	4.00	3.15	2.76	2.53	2.37	2.25	2.17	2.10	2.04	1.99	1.95	1.92	1.89	1.86	1.84	1.82	1.80	1.78	1.76
70	3.98	3.13	2.74	2.50	2.35	2.23	2.14	2.07	2.02	1.97	1.93	1.89	1.86	1.84	1.81	1.79	1.77	1.75	1.74
80	3.96	3.11	2.72	2.49	2.33	2.21	2.13	2.06	2.00	1.95	1.91	1.88	1.84	1.82	1.79	1.77	1.75	1.73	1.72
90	3.95	3.10	2.71	2.47	2.32	2.20	2.11	2.04	1.99	1.94	1.90	1.86	1.83	1.80	1.78	1.76	1.74	1.72	1.70
100	3.94	3.09	2.70	2.46	2.31	2.19	2.10	2.03	1.97	1.93	1.89	1.85	1.82	1.79	1.77	1.75	1.73	1.71	1.69
110	3.93	3.08	2.69	2.45	2.30	2.18	2.09	2.02	1.97	1.92	1.88	1.84	1.81	1.78	1.76	1.74	1.72	1.70	1.58
120	3.92	3.07	2.68	2.45	2.29	2.18	2.09	2.02	1.95	1.91	1.87	1.83	1.80	1.78	1.75	1.73	1.71	1.69	1.57
10000	3.84	3.00	2.61	2.37	2.21	2.10	2.01	1.94	1.88	1.83	1.79	1.75	1.72	1.69	1.67	1.64	1.62	1.60	1.59

Table B.3 (Continued)

Critical values of the F distribution at 0.05 significance level

	Degrees of freedom, v_1																				
Degrees of freedom, v_2	20	21	22	23	24	25	26	27	28	29	30	40	50	60	70	80	90	100	110	120	10000
1	248.0	248.3	248.6	248.8	249.1	249.3	249.5	249.6	249.8	250.0	250.1	251.1	251.8	252.2	252.5	252.7	252.9	253.0	253.2	253.3	254.3
2	19.45	19.45	19.45	19.45	19.45	19.46	19.46	19.46	19.46	19.46	19.46	19.47	19.47	19.48	19.48	19.48	19.48	19.49	19.49	19.49	19.50
3	8.66	8.65	8.65	8.64	8.64	8.63	8.63	8.63	8.62	8.62	8.62	8.59	8.58	8.57	8.57	8.56	8.56	8.55	8.55	8.55	8.53
4	5.80	5.79	5.79	5.78	5.77	5.77	5.76	5.76	5.75	5.75	5.75	5.72	5.70	5.69	5.68	5.67	5.67	5.66	5.66	5.66	5.63
5	4.56	4.55	4.54	4.53	4.53	4.52	4.52	4.51	4.50	4.50	4.50	4.46	4.44	4.43	4.42	4.41	4.41	4.41	4.40	4.40	4.37
6	3.87	3.86	3.86	3.85	3.84	3.83	3.83	3.82	3.82	3.81	3.81	3.77	3.75	3.74	3.73	3.72	3.72	3.71	3.71	3.70	3.67
7	3.44	3.43	3.43	3.42	3.41	3.40	3.40	3.39	3.39	3.38	3.38	3.34	3.32	3.30	3.29	3.29	3.29	3.27	3.27	3.27	3.23
8	3.15	3.14	3.13	3.12	3.12	3.10	3.10	3.10	3.09	3.08	3.08	3.04	3.02	3.01	2.99	2.99	2.98	2.97	2.97	2.97	2.93
9	2.94	2.93	2.92	2.91	2.90	2.89	2.89	2.88	2.87	2.87	2.86	2.83	2.80	2.79	2.78	2.77	2.76	2.76	2.75	2.75	2.71
10	2.77	2.76	2.75	2.75	2.74	2.72	2.72	2.72	2.71	2.70	2.70	2.66	2.64	2.62	2.61	2.60	2.59	2.59	2.58	2.58	2.54
11	2.65	2.64	2.63	2.62	2.61	2.59	2.59	2.59	2.58	2.58	2.57	2.53	2.51	2.49	2.48	2.48	2.47	2.46	2.45	2.45	2.41
12	2.54	2.53	2.52	2.51	2.51	2.50	2.49	2.48	2.48	2.47	2.47	2.43	2.40	2.38	2.37	2.37	2.36	2.35	2.34	2.34	2.30
13	2.46	2.45	2.44	2.43	2.42	2.41	2.41	2.40	2.39	2.39	2.38	2.34	2.31	2.30	2.28	2.28	2.27	2.26	2.26	2.25	2.21
14	2.39	2.38	2.37	2.36	2.35	2.34	2.33	2.33	2.32	2.32	2.31	2.27	2.24	2.22	2.21	2.21	2.20	2.19	2.18	2.18	2.13
15	2.33	2.32	2.31	2.30	2.29	2.28	2.27	2.27	2.26	2.25	2.25	2.20	2.18	2.16	2.15	2.15	2.14	2.13	2.12	2.11	2.07
16	2.28	2.26	2.25	2.24	2.24	2.23	2.22	2.21	2.21	2.20	2.20	2.15	2.12	2.11	2.09	2.09	2.08	2.07	2.06	2.06	2.01
17	2.23	2.22	2.21	2.20	2.19	2.18	2.17	2.17	2.16	2.15	2.15	2.10	2.08	2.06	2.05	2.05	2.03	2.02	2.02	2.01	1.96

18	1.92	1.97	1.97	1.98	1.98	1.99	2.00	2.02	2.04	2.06	2.11	2.11	2.12	2.13	2.13	2.14	2.15	2.16	2.17	2.18	2.19
19	1.88	1.93	1.93	1.94	1.95	1.96	1.97	1.98	2.00	2.03	2.07	2.08	2.08	2.09	2.10	2.11	2.11	2.12	2.13	2.14	2.16
20	1.84	1.90	1.90	1.91	1.91	1.92	1.93	1.95	1.97	1.99	2.04	2.05	2.05	2.06	2.07	2.07	2.08	2.09	2.10	2.11	2.12
21	1.81	1.87	1.87	1.88	1.88	1.89	1.90	1.92	1.94	1.96	2.01	2.02	2.02	2.03	2.04	2.05	2.05	2.06	2.07	2.08	2.10
22	1.78	1.84	1.84	1.85	1.86	1.86	1.88	1.89	1.91	1.94	1.98	1.99	2.00	2.00	2.01	2.02	2.03	2.04	2.05	2.06	2.07
23	1.76	1.81	1.82	1.82	1.83	1.84	1.85	1.86	1.88	1.91	1.96	1.97	1.97	1.98	1.99	2.00	2.01	2.01	2.02	2.04	2.05
24	1.73	1.79	1.79	1.80	1.81	1.82	1.83	1.84	1.86	1.89	1.94	1.95	1.95	1.96	1.97	1.97	1.98	1.99	2.00	2.01	2.03
25	1.71	1.77	1.77	1.78	1.79	1.80	1.81	1.82	1.84	1.87	1.92	1.93	1.93	1.94	1.95	1.96	1.96	1.97	1.98	2.00	2.01
26	1.69	1.75	1.75	1.76	1.77	1.78	1.79	1.80	1.82	1.85	1.90	1.91	1.91	1.92	1.93	1.94	1.95	1.96	1.97	1.98	1.99
27	1.67	1.73	1.74	1.74	1.75	1.76	1.77	1.79	1.81	1.84	1.88	1.89	1.90	1.90	1.91	1.92	1.93	1.94	1.95	1.96	1.97
28	1.65	1.71	1.72	1.73	1.73	1.74	1.75	1.77	1.79	1.82	1.87	1.88	1.88	1.89	1.90	1.91	1.91	1.92	1.93	1.95	1.96
29	1.64	1.70	1.70	1.71	1.72	1.73	1.74	1.75	1.77	1.81	1.85	1.86	1.87	1.88	1.88	1.89	1.90	1.91	1.92	1.93	1.94
30	1.62	1.68	1.69	1.70	1.70	1.71	1.72	1.74	1.76	1.79	1.84	1.85	1.85	1.86	1.87	1.88	1.89	1.90	1.91	1.92	1.93
40	1.51	1.58	1.58	1.59	1.60	1.61	1.62	1.64	1.66	1.69	1.74	1.75	1.76	1.77	1.77	1.78	1.79	1.80	1.81	1.83	1.84
50	1.44	1.51	1.52	1.52	1.53	1.54	1.56	1.58	1.60	1.63	1.69	1.69	1.70	1.71	1.72	1.73	1.74	1.75	1.76	1.77	1.78
60	1.39	1.47	1.47	1.48	1.49	1.50	1.52	1.53	1.56	1.59	1.65	1.66	1.66	1.67	1.68	1.69	1.70	1.71	1.72	1.73	1.75
70	1.35	1.44	1.44	1.45	1.46	1.47	1.49	1.50	1.53	1.57	1.62	1.63	1.64	1.65	1.65	1.66	1.67	1.68	1.70	1.71	1.72
80	1.33	1.41	1.42	1.43	1.44	1.45	1.46	1.48	1.51	1.54	1.61	1.61	1.62	1.63	1.63	1.64	1.65	1.67	1.68	1.69	1.70
90	1.30	1.39	1.40	1.41	1.42	1.43	1.44	1.46	1.49	1.53	1.59	1.59	1.60	1.61	1.62	1.63	1.64	1.65	1.66	1.67	1.69
100	1.28	1.38	1.38	1.39	1.40	1.41	1.43	1.45	1.48	1.52	1.57	1.58	1.59	1.60	1.61	1.62	1.63	1.64	1.65	1.66	1.68
110	1.27	1.36	1.37	1.38	1.39	1.40	1.42	1.44	1.47	1.50	1.56	1.57	1.58	1.59	1.60	1.61	1.62	1.63	1.64	1.65	1.67
120	1.26	1.35	1.36	1.37	1.38	1.39	1.41	1.43	1.46	1.50	1.55	1.56	1.57	1.58	1.59	1.60	1.61	1.62	1.63	1.64	1.66
10000	1.00	1.22	1.23	1.25	1.26	1.28	1.29	1.32	1.35	1.40	1.46	1.47	1.48	1.49	1.50	1.51	1.52	1.53	1.54	1.56	1.57

Table B.3 (Continued)

Critical values of the F distribution at 0.01 significance level

ν_2 \ ν_1	1	2	3	4	5	6	7	8	9	10	11	12	13	14	15	16	17	18
1	4052	5000	5403	5625	5764	5859	5928	5981	6022	6056	6083	6106	6126	6143	6157	6170	6181	6192
2	98.50	99.00	99.17	99.25	99.30	99.33	99.36	99.37	99.39	99.40	99.41	99.42	99.42	99.43	99.43	99.44	99.44	99.44
3	34.12	30.82	29.46	28.71	28.24	27.91	27.67	27.49	27.35	27.23	27.13	27.05	26.98	26.92	26.87	26.83	26.79	26.75
4	21.20	18.00	16.69	15.98	15.52	15.21	14.98	14.80	14.66	14.55	14.45	14.37	14.31	14.25	14.20	14.15	14.11	14.08
5	16.26	13.27	12.06	11.39	10.97	10.67	10.46	10.29	10.16	10.05	9.96	9.89	9.82	9.77	9.72	9.68	9.64	9.61
6	13.75	10.92	9.78	9.15	8.75	8.47	8.26	8.10	7.98	7.87	7.79	7.72	7.66	7.60	7.56	7.52	7.48	7.45
7	12.25	9.55	8.45	7.85	7.46	7.19	6.99	6.84	6.72	6.62	6.54	6.47	6.41	6.36	6.31	6.28	6.24	6.21
8	11.26	8.65	7.59	7.01	6.63	6.37	6.18	6.03	5.91	5.81	5.73	5.67	5.61	5.56	5.52	5.48	5.44	5.41
9	10.56	8.02	6.99	6.42	6.06	5.80	5.61	5.47	5.35	5.26	5.18	5.11	5.05	5.01	4.96	4.92	4.89	4.86
10	10.04	7.56	6.55	5.99	5.64	5.39	5.20	5.06	4.94	4.85	4.77	4.71	4.65	4.60	4.56	4.52	4.49	4.46
11	9.65	7.21	6.22	5.67	5.32	5.07	4.89	4.74	4.63	4.54	4.46	4.40	4.34	4.29	4.25	4.21	4.18	4.15
12	9.33	6.93	5.95	5.41	5.06	4.82	4.64	4.50	4.39	4.30	4.22	4.16	4.10	4.05	4.01	3.97	3.94	3.91
13	9.07	6.70	5.74	5.21	4.86	4.62	4.44	4.30	4.19	4.10	4.02	3.96	3.91	3.86	3.82	3.78	3.75	3.72
14	8.86	6.51	5.56	5.04	4.69	4.46	4.28	4.14	4.03	3.94	3.86	3.80	3.75	3.70	3.66	3.62	3.59	3.56
15	8.68	6.36	5.42	4.89	4.56	4.32	4.14	4.00	3.89	3.80	3.73	3.67	3.61	3.56	3.52	3.49	3.45	3.42
16	8.53	6.23	5.29	4.77	4.44	4.20	4.03	3.89	3.78	3.69	3.62	3.55	3.50	3.45	3.41	3.37	3.34	3.31
17	8.40	6.11	5.18	4.67	4.34	4.10	3.93	3.79	3.68	3.59	3.52	3.46	3.40	3.35	3.31	3.27	3.24	3.21

Degrees of freedom, ν_1

Degrees of freedom, ν_2

18	8.29	6.01	5.09	4.58	4.25	4.01	3.84	3.71	3.60	3.51	3.43	3.37	3.32	3.27	3.23	3.19	3.16	3.13
19	8.18	5.93	5.01	4.50	4.17	3.94	3.77	3.63	3.52	3.43	3.36	3.30	3.24	3.19	3.15	3.12	3.08	3.05
20	8.10	5.85	4.94	4.43	4.10	3.87	3.70	3.56	3.46	3.37	3.29	3.23	3.18	3.13	3.09	3.05	3.02	2.99
21	8.02	5.78	4.87	4.37	4.04	3.81	3.64	3.51	3.40	3.31	3.24	3.17	3.12	3.07	3.03	2.99	2.96	2.93
22	7.95	5.72	4.82	4.31	3.99	3.76	3.59	3.45	3.35	3.26	3.18	3.12	3.07	3.02	2.98	2.94	2.91	2.88
23	7.88	5.66	4.76	4.26	3.94	3.71	3.54	3.41	3.30	3.21	3.14	3.07	3.02	2.97	2.93	2.89	2.86	2.83
24	7.82	5.61	4.72	4.22	3.90	3.67	3.50	3.36	3.26	3.17	3.09	3.03	2.98	2.93	2.89	2.85	2.82	2.79
25	7.77	5.57	4.68	4.18	3.85	3.63	3.46	3.32	3.22	3.13	3.06	2.99	2.94	2.89	2.85	2.81	2.78	2.75
26	7.72	5.53	4.64	4.14	3.82	3.59	3.42	3.29	3.18	3.09	3.02	2.96	2.90	2.86	2.81	2.78	2.75	2.72
27	7.68	5.49	4.60	4.11	3.78	3.56	3.39	3.26	3.15	3.06	2.99	2.93	2.87	2.82	2.78	2.75	2.71	2.68
28	7.64	5.45	4.57	4.07	3.75	3.53	3.36	3.23	3.12	3.03	2.96	2.90	2.84	2.79	2.75	2.72	2.68	2.65
29	7.60	5.42	4.54	4.04	3.73	3.50	3.33	3.20	3.09	3.00	2.93	2.87	2.81	2.77	2.73	2.69	2.66	2.63
30	7.56	5.39	4.51	4.02	3.70	3.47	3.30	3.17	3.07	2.98	2.91	2.84	2.79	2.74	2.70	2.66	2.63	2.50
40	7.31	5.18	4.31	3.83	3.51	3.29	3.12	2.99	2.89	2.80	2.73	2.66	2.61	2.56	2.52	2.48	2.45	2.42
50	7.17	5.06	4.20	3.72	3.41	3.19	3.02	2.89	2.78	2.70	2.63	2.56	2.51	2.46	2.42	2.38	2.35	2.32
60	7.08	4.98	4.13	3.65	3.34	3.12	2.95	2.82	2.72	2.63	2.56	2.50	2.44	2.39	2.35	2.31	2.28	2.25
70	7.01	4.92	4.07	3.60	3.29	3.07	2.91	2.78	2.67	2.59	2.51	2.45	2.40	2.35	2.31	2.27	2.23	2.20
80	6.96	4.88	4.04	3.56	3.26	3.04	2.87	2.74	2.64	2.55	2.48	2.42	2.36	2.31	2.27	2.23	2.20	2.17
90	6.93	4.85	4.01	3.53	3.23	3.01	2.84	2.72	2.61	2.52	2.45	2.39	2.33	2.29	2.24	2.21	2.17	2.14
100	6.90	4.82	3.98	3.51	3.21	2.99	2.82	2.69	2.59	2.50	2.43	2.37	2.31	2.27	2.22	2.19	2.15	2.12
110	6.87	4.80	3.96	3.49	3.19	2.97	2.81	2.68	2.57	2.49	2.41	2.35	2.30	2.25	2.21	2.17	2.13	2.10
120	6.85	4.79	3.95	3.48	3.17	2.96	2.79	2.66	2.56	2.47	2.40	2.34	2.28	2.23	2.19	2.15	2.12	2.09
10000	6.64	4.61	3.78	3.32	3.02	2.80	2.64	2.51	2.41	2.32	2.25	2.19	2.13	2.08	2.04	2.00	1.97	1.94

Table B.3 (Continued)

Critical values of the F distribution at 0.01 significance level

	Degrees of freedom, v_1																					
Degrees of freedom, v_2	19	20	21	22	23	24	25	26	27	28	29	30	40	50	60	70	80	90	100	110	120	10000
1	6201	6209	6216	6223	6229	6235	6240	6245	6249	6253	6257	6261	6287	6303	6313	6321	6326	6331	6334	6337	6339	6366
2	99.45	99.45	99.45	99.45	99.46	99.46	99.46	99.46	99.46	99.46	99.46	99.47	99.47	99.48	99.48	99.48	99.49	99.49	99.49	99.49	99.49	99.50
3	26.72	26.69	26.66	26.64	26.62	26.60	26.58	26.56	26.55	26.53	26.52	26.50	26.41	26.35	26.32	26.29	26.27	26.25	26.24	26.23	26.22	26.13
4	14.05	14.02	13.99	13.97	13.95	13.93	13.91	13.89	13.88	13.86	13.85	13.84	13.75	13.69	13.65	13.63	13.61	13.59	13.58	13.57	13.56	13.46
5	9.58	9.55	9.53	9.51	9.49	9.47	9.45	9.43	9.42	9.40	9.39	9.38	9.29	9.24	9.20	9.18	9.16	9.14	9.13	9.12	9.11	9.02
6	7.42	7.40	7.37	7.35	7.33	7.31	7.30	7.28	7.27	7.25	7.24	7.23	7.14	7.09	7.06	7.03	7.01	7.00	6.99	6.98	6.97	6.88
7	6.18	6.16	6.13	6.11	6.09	6.07	6.06	6.04	6.03	6.02	6.00	5.99	5.91	5.86	5.82	5.80	5.78	5.77	5.75	5.75	5.74	5.65
8	5.38	5.36	5.34	5.32	5.30	5.28	5.26	5.25	5.23	5.22	5.21	5.20	5.12	5.07	5.03	5.01	4.99	4.97	4.96	4.95	4.95	4.86
9	4.83	4.81	4.79	4.77	4.75	4.73	4.71	4.70	4.68	4.67	4.66	4.65	4.57	4.52	4.48	4.46	4.44	4.43	4.41	4.41	4.40	4.31
10	4.43	4.41	4.38	4.36	4.34	4.33	4.31	4.30	4.28	4.27	4.26	4.25	4.17	4.12	4.08	4.06	4.04	4.03	4.01	4.00	4.00	3.91
11	4.12	4.10	4.08	4.06	4.04	4.02	4.01	3.99	3.98	3.96	3.95	3.94	3.86	3.81	3.78	3.75	3.73	3.72	3.71	3.70	3.69	3.60
12	3.88	3.86	3.84	3.82	3.80	3.78	3.76	3.75	3.74	3.72	3.71	3.70	3.62	3.57	3.54	3.51	3.49	3.48	3.47	3.46	3.45	3.36
13	3.69	3.66	3.64	3.62	3.60	3.59	3.57	3.56	3.54	3.53	3.52	3.51	3.43	3.38	3.34	3.32	3.30	3.28	3.27	3.26	3.25	3.17
14	3.53	3.51	3.48	3.46	3.44	3.43	3.41	3.40	3.38	3.37	3.36	3.35	3.27	3.22	3.18	3.16	3.14	3.12	3.11	3.10	3.09	3.01
15	3.40	3.37	3.35	3.33	3.31	3.29	3.28	3.26	3.25	3.24	3.23	3.21	3.13	3.08	3.05	3.02	3.00	2.99	2.98	2.97	2.96	2.87
16	3.28	3.26	3.24	3.22	3.20	3.18	3.16	3.15	3.14	3.12	3.11	3.10	3.02	2.97	2.93	2.91	2.89	2.87	2.86	2.85	2.84	2.75
17	3.19	3.16	3.14	3.12	3.10	3.08	3.07	3.05	3.04	3.03	3.01	3.00	2.92	2.87	2.83	2.81	2.79	2.78	2.76	2.75	2.75	2.65

18	2.57	2.66	2.67	2.68	2.69	2.70	2.72	2.75	2.78	2.84	2.92	2.93	2.94	2.95	2.97	2.98	3.00	3.02	3.03	3.05	3.08	3.10
19	2.49	2.58	2.59	2.60	2.61	2.63	2.65	2.67	2.71	2.76	2.84	2.86	2.87	2.88	2.89	2.91	2.92	2.94	2.96	2.98	3.00	3.03
20	2.42	2.52	2.53	2.54	2.55	2.56	2.58	2.61	2.64	2.69	2.78	2.79	2.80	2.81	2.83	2.84	2.86	2.88	2.90	2.92	2.94	2.96
21	2.36	2.46	2.47	2.48	2.49	2.50	2.52	2.55	2.58	2.64	2.72	2.73	2.74	2.76	2.77	2.79	2.80	2.82	2.84	2.86	2.88	2.90
22	2.31	2.40	2.41	2.42	2.43	2.45	2.47	2.50	2.53	2.58	2.67	2.68	2.69	2.70	2.72	2.73	2.75	2.77	2.78	2.81	2.83	2.85
23	2.26	2.35	2.36	2.37	2.39	2.40	2.42	2.45	2.48	2.54	2.62	2.63	2.64	2.66	2.67	2.69	2.70	2.72	2.74	2.76	2.78	2.80
24	2.21	2.31	2.32	2.33	2.34	2.36	2.38	2.40	2.44	2.49	2.58	2.59	2.60	2.61	2.63	2.64	2.66	2.68	2.70	2.72	2.74	2.76
25	2.17	2.27	2.28	2.29	2.30	2.32	2.34	2.36	2.40	2.45	2.54	2.55	2.56	2.58	2.59	2.60	2.62	2.64	2.66	2.68	2.70	2.72
26	2.13	2.23	2.24	2.25	2.26	2.28	2.30	2.33	2.36	2.42	2.50	2.51	2.53	2.54	2.55	2.57	2.58	2.60	2.62	2.64	2.66	2.69
27	2.10	2.20	2.21	2.22	2.23	2.25	2.27	2.29	2.33	2.38	2.47	2.48	2.49	2.51	2.52	2.54	2.55	2.57	2.59	2.61	2.63	2.66
28	2.07	2.17	2.18	2.19	2.20	2.22	2.24	2.26	2.30	2.35	2.44	2.45	2.46	2.48	2.49	2.51	2.52	2.54	2.56	2.58	2.60	2.63
29	2.04	2.14	2.15	2.16	2.17	2.19	2.21	2.23	2.27	2.33	2.41	2.42	2.44	2.45	2.46	2.48	2.49	2.51	2.53	2.55	2.57	2.60
30	2.01	2.11	2.12	2.13	2.14	2.16	2.18	2.21	2.25	2.30	2.39	2.40	2.41	2.42	2.44	2.45	2.47	2.49	2.51	2.53	2.55	2.57
40	1.81	1.92	1.93	1.94	1.95	1.97	1.99	2.02	2.06	2.11	2.20	2.22	2.23	2.24	2.26	2.27	2.29	2.31	2.33	2.35	2.37	2.39
50	1.68	1.80	1.81	1.82	1.84	1.86	1.88	1.91	1.95	2.01	2.10	2.11	2.12	2.14	2.15	2.17	2.18	2.20	2.22	2.24	2.27	2.29
60	1.60	1.73	1.74	1.75	1.76	1.78	1.81	1.84	1.88	1.94	2.03	2.04	2.05	2.07	2.08	2.10	2.12	2.13	2.15	2.17	2.20	2.22
70	1.54	1.67	1.68	1.70	1.71	1.73	1.75	1.78	1.83	1.89	1.98	1.99	2.01	2.02	2.03	2.05	2.07	2.09	2.11	2.13	2.15	2.18
80	1.50	1.63	1.64	1.65	1.67	1.69	1.71	1.75	1.79	1.85	1.94	1.96	1.97	1.98	2.00	2.01	2.03	2.05	2.07	2.09	2.12	2.14
90	1.46	1.60	1.61	1.62	1.64	1.66	1.68	1.72	1.76	1.82	1.92	1.93	1.94	1.96	1.97	1.99	2.00	2.02	2.04	2.06	2.09	2.11
100	1.43	1.57	1.58	1.60	1.61	1.63	1.66	1.69	1.74	1.80	1.89	1.91	1.92	1.93	1.95	1.97	1.98	2.00	2.02	2.04	2.07	2.09
110	1.40	1.55	1.56	1.58	1.59	1.61	1.64	1.67	1.72	1.78	1.88	1.89	1.90	1.92	1.93	1.95	1.96	1.98	2.00	2.03	2.05	2.07
120	1.38	1.53	1.55	1.56	1.58	1.60	1.62	1.66	1.70	1.76	1.86	1.87	1.89	1.90	1.92	1.93	1.95	1.97	1.99	2.01	2.03	2.06
10000	1.00	1.33	1.34	1.36	1.38	1.41	1.44	1.48	1.53	1.59	1.70	1.71	1.73	1.74	1.76	1.77	1.79	1.81	1.83	1.86	1.88	1.91

Table B.3 (Continued)

Critical values of the F distribution at 0.005 significance level

ν_2 \ ν_1	1	2	3	4	5	6	7	8	9	10	11	12	13	14	15	16	17	18	19
1	16211	20000	21615	22500	23056	23437	23715	23925	24091	24224	24334	24426	24505	24572	24630	24681	24727	24767	24803
2	198.5	199.0	199.2	199.2	199.3	199.3	199.4	199.4	199.4	199.4	199.4	199.4	199.4	199.4	199.4	199.4	199.4	199.4	199.4
3	55.55	49.80	47.47	46.19	45.39	44.84	44.43	44.13	43.88	43.69	43.52	43.39	43.27	43.17	43.08	43.01	42.94	42.88	42.83
4	31.33	26.28	24.26	23.15	22.46	21.97	21.62	21.35	21.14	20.97	20.82	20.70	20.60	20.51	20.44	20.37	20.31	20.26	20.21
5	22.78	18.31	16.53	15.56	14.94	14.51	14.20	13.96	13.77	13.62	13.49	13.38	13.29	13.21	13.15	13.09	13.03	12.98	12.94
6	18.63	14.54	12.92	12.03	11.46	11.07	10.79	10.57	10.39	10.25	10.13	10.03	9.95	9.88	9.81	9.76	9.71	9.66	9.62
7	16.24	12.40	10.88	10.05	9.52	9.16	8.89	8.68	8.51	8.38	8.27	8.18	8.10	8.03	7.97	7.91	7.87	7.83	7.79
8	14.69	11.04	9.60	8.81	8.30	7.95	7.69	7.50	7.34	7.21	7.10	7.01	6.94	6.87	6.81	6.76	6.72	6.68	6.64
9	13.61	10.11	8.72	7.96	7.47	7.13	6.88	6.69	6.54	6.42	6.31	6.23	6.15	6.09	6.03	5.98	5.94	5.90	5.86
10	12.83	9.43	8.08	7.34	6.87	6.54	6.30	6.12	5.97	5.85	5.75	5.66	5.59	5.53	5.47	5.42	5.38	5.34	5.31
11	12.23	8.91	7.60	6.88	6.42	6.10	5.86	5.68	5.54	5.42	5.32	5.24	5.16	5.10	5.05	5.00	4.96	4.92	4.89
12	11.75	8.51	7.23	6.52	6.07	5.76	5.52	5.35	5.20	5.09	4.99	4.91	4.84	4.77	4.72	4.67	4.63	4.59	4.56
13	11.37	8.19	6.93	6.23	5.79	5.48	5.25	5.08	4.94	4.82	4.72	4.64	4.57	4.51	4.46	4.41	4.37	4.33	4.30
14	11.06	7.92	6.68	6.00	5.56	5.26	5.03	4.86	4.72	4.60	4.51	4.43	4.36	4.30	4.25	4.20	4.16	4.12	4.09
15	10.80	7.70	6.48	5.80	5.37	5.07	4.85	4.67	4.54	4.42	4.33	4.25	4.18	4.12	4.07	4.02	3.98	3.95	3.91
16	10.58	7.51	6.30	5.64	5.21	4.91	4.69	4.52	4.38	4.27	4.18	4.10	4.03	3.97	3.92	3.87	3.83	3.80	3.76
17	10.38	7.35	6.16	5.50	5.07	4.78	4.56	4.39	4.25	4.14	4.05	3.97	3.90	3.84	3.79	3.75	3.71	3.67	3.64

Degrees of freedom, ν_1

Degrees of freedom, ν_2

18	10.22	7.21	6.03	5.37	4.96	4.66	4.44	4.28	4.14	4.03	3.94	3.86	3.79	3.73	3.68	3.64	3.60	3.56	3.53
19	10.07	7.09	5.92	5.27	4.85	4.56	4.34	4.18	4.04	3.93	3.84	3.76	3.70	3.64	3.59	3.54	3.50	3.46	3.43
20	9.94	6.99	5.82	5.17	4.76	4.47	4.26	4.09	3.96	3.85	3.76	3.68	3.61	3.55	3.50	3.46	3.42	3.38	3.35
21	9.83	6.89	5.73	5.09	4.68	4.39	4.18	4.01	3.88	3.77	3.68	3.60	3.54	3.48	3.43	3.38	3.34	3.31	3.27
22	9.73	6.81	5.65	5.02	4.61	4.32	4.11	3.94	3.81	3.70	3.61	3.54	3.47	3.41	3.36	3.31	3.27	3.24	3.21
23	9.63	6.73	5.58	4.95	4.54	4.26	4.05	3.88	3.75	3.64	3.55	3.47	3.41	3.35	3.30	3.25	3.21	3.18	3.15
24	9.55	6.66	5.52	4.89	4.49	4.20	3.99	3.83	3.69	3.59	3.50	3.42	3.35	3.30	3.25	3.20	3.16	3.12	3.09
25	9.48	6.60	5.46	4.84	4.43	4.15	3.94	3.78	3.64	3.54	3.45	3.37	3.30	3.25	3.20	3.15	3.11	3.08	3.04
26	9.41	6.54	5.41	4.79	4.38	4.10	3.89	3.73	3.60	3.49	3.40	3.33	3.26	3.20	3.15	3.11	3.07	3.03	3.00
27	9.34	6.49	5.36	4.74	4.34	4.06	3.85	3.69	3.56	3.45	3.36	3.28	3.22	3.16	3.11	3.07	3.03	2.99	2.96
28	9.28	6.44	5.32	4.70	4.30	4.02	3.81	3.65	3.52	3.41	3.32	3.25	3.18	3.12	3.07	3.03	2.99	2.95	2.92
29	9.23	6.40	5.28	4.66	4.26	3.98	3.77	3.61	3.48	3.38	3.29	3.21	3.15	3.09	3.04	2.99	2.95	2.92	2.88
30	9.18	6.35	5.24	4.62	4.23	3.95	3.74	3.58	3.45	3.34	3.25	3.18	3.11	3.06	3.01	2.96	2.92	2.89	2.85
40	8.83	6.07	4.98	4.37	3.99	3.71	3.51	3.35	3.22	3.12	3.03	2.95	2.89	2.83	2.78	2.74	2.70	2.66	2.63
50	8.63	5.90	4.83	4.23	3.85	3.58	3.38	3.22	3.09	2.99	2.90	2.82	2.76	2.70	2.65	2.61	2.57	2.53	2.50
60	8.49	5.79	4.73	4.14	3.76	3.49	3.29	3.13	3.01	2.90	2.82	2.74	2.68	2.62	2.57	2.53	2.49	2.45	2.42
70	8.40	5.72	4.66	4.08	3.70	3.43	3.23	3.08	2.95	2.85	2.76	2.68	2.62	2.56	2.51	2.47	2.43	2.39	2.36
80	8.33	5.67	4.61	4.03	3.65	3.39	3.19	3.03	2.91	2.80	2.72	2.64	2.58	2.52	2.47	2.43	2.39	2.35	2.32
90	8.28	5.62	4.57	3.99	3.62	3.35	3.15	3.00	2.87	2.77	2.68	2.61	2.54	2.49	2.44	2.39	2.35	2.32	2.28
100	8.24	5.59	4.54	3.96	3.59	3.33	3.13	2.97	2.85	2.74	2.66	2.58	2.52	2.46	2.41	2.37	2.33	2.29	2.26
110	8.21	5.56	4.52	3.94	3.57	3.30	3.11	2.95	2.83	2.72	2.64	2.56	2.50	2.44	2.39	2.35	2.31	2.27	2.24
120	8.18	5.54	4.50	3.92	3.55	3.28	3.09	2.93	2.81	2.71	2.62	2.54	2.48	2.42	2.37	2.33	2.29	2.25	2.22
10000	7.88	5.30	4.28	3.72	3.35	3.09	2.90	2.75	2.62	2.52	2.43	2.36	2.30	2.24	2.19	2.14	2.10	2.07	2.03

Table B.3 (Continued)

Critical values of the F distribution at 0.005 significance level

v_2	Degrees of freedom, v_1																				
	20	21	22	23	24	25	26	27	28	29	30	40	50	60	70	80	90	100	110	120	10000
1	24836	24866	24892	24917	24940	24960	24980	24997	25014	25029	25044	25148	25211	25253	25283	25306	25323	25337	25349	25359	25463
2	199.4	199.5	199.5	199.5	199.5	199.5	199.5	199.5	199.5	199.5	199.5	199.5	199.5	199.5	199.5	199.5	199.5	199.5	199.5	199.5	199.5
3	42.78	42.73	42.69	42.66	42.62	42.59	42.56	42.54	42.51	42.49	42.47	42.31	42.21	42.15	42.10	42.07	42.04	42.02	42.00	41.99	41.83
4	20.17	20.13	20.09	20.06	20.03	20.00	19.98	19.95	19.93	19.91	19.89	19.75	19.67	19.61	19.57	19.54	19.52	19.50	19.48	19.47	19.33
5	12.90	12.87	12.84	12.81	12.78	12.76	12.73	12.71	12.69	12.67	12.66	12.53	12.45	12.40	12.37	12.34	12.32	12.30	12.29	12.27	12.15
6	9.59	9.56	9.53	9.50	9.47	9.45	9.43	9.41	9.39	9.37	9.36	9.24	9.17	9.12	9.09	9.06	9.04	9.03	9.01	9.00	8.88
7	7.75	7.72	7.69	7.67	7.64	7.62	7.60	7.58	7.57	7.55	7.53	7.42	7.35	7.31	7.28	7.25	7.23	7.22	7.20	7.19	7.08
8	6.61	6.58	6.55	6.53	6.50	6.48	6.46	6.44	6.43	6.41	6.40	6.29	6.22	6.18	6.15	6.12	6.10	6.09	6.08	6.06	5.95
9	5.83	5.80	5.78	5.75	5.73	5.71	5.69	5.67	5.65	5.64	5.62	5.52	5.45	5.41	5.38	5.36	5.34	5.32	5.31	5.30	5.19
10	5.27	5.25	5.22	5.20	5.17	5.15	5.13	5.12	5.10	5.08	5.07	4.97	4.90	4.86	4.83	4.80	4.79	4.77	4.76	4.75	4.64
11	4.86	4.83	4.80	4.78	4.76	4.74	4.72	4.70	4.68	4.67	4.65	4.55	4.49	4.45	4.41	4.39	4.37	4.36	4.35	4.34	4.23
12	4.53	4.50	4.48	4.45	4.43	4.41	4.39	4.38	4.36	4.34	4.33	4.23	4.17	4.12	4.09	4.07	4.05	4.04	4.02	4.01	3.91
13	4.27	4.24	4.22	4.19	4.17	4.15	4.13	4.12	4.10	4.09	4.07	3.97	3.91	3.87	3.84	3.81	3.79	3.78	3.77	3.76	3.65
14	4.06	4.03	4.01	3.98	3.96	3.94	3.92	3.91	3.89	3.88	3.86	3.76	3.70	3.66	3.62	3.60	3.58	3.57	3.56	3.55	3.44
15	3.88	3.86	3.83	3.81	3.79	3.77	3.75	3.73	3.72	3.70	3.69	3.58	3.52	3.48	3.45	3.43	3.41	3.39	3.38	3.37	3.26
16	3.73	3.71	3.68	3.66	3.64	3.62	3.60	3.58	3.57	3.55	3.54	3.44	3.37	3.33	3.30	3.28	3.26	3.25	3.23	3.22	3.11
17	3.61	3.58	3.56	3.53	3.51	3.49	3.47	3.46	3.44	3.43	3.41	3.31	3.25	3.21	3.18	3.15	3.13	3.12	3.11	3.10	2.99

Degrees of freedom, v_2

18	2.87	2.99	3.00	3.01	3.02	3.04	3.07	3.10	3.14	3.20	3.30	3.32	3.33	3.35	3.36	3.38	3.40	3.42	3.45	3.47	3.50
19	2.78	2.89	2.90	2.91	2.93	2.95	2.97	3.00	3.04	3.11	3.21	3.22	3.24	3.25	3.27	3.29	3.31	3.33	3.35	3.37	3.40
20	2.69	2.81	2.82	2.83	2.84	2.86	2.88	2.92	2.96	3.02	3.12	3.14	3.15	3.17	3.18	3.20	3.22	3.24	3.27	3.29	3.32
21	2.62	2.73	2.74	2.75	2.77	2.79	2.81	2.84	2.88	2.95	3.05	3.06	3.08	3.09	3.11	3.13	3.15	3.17	3.19	3.22	3.24
22	2.55	2.66	2.67	2.69	2.70	2.72	2.74	2.77	2.82	2.88	2.98	3.00	3.01	3.03	3.04	3.06	3.08	3.10	3.12	3.15	3.18
23	2.49	2.60	2.61	2.62	2.64	2.66	2.68	2.71	2.76	2.82	2.92	2.94	2.95	2.97	2.98	3.00	3.02	3.04	3.06	3.09	3.12
24	2.45	2.55	2.56	2.57	2.58	2.60	2.63	2.66	2.70	2.77	2.87	2.88	2.90	2.91	2.93	2.95	2.97	2.99	3.01	3.04	3.06
25	2.38	2.50	2.51	2.52	2.53	2.55	2.58	2.61	2.65	2.72	2.82	2.83	2.85	2.86	2.88	2.90	2.92	2.94	2.96	2.99	3.01
26	2.33	2.45	2.46	2.47	2.49	2.51	2.53	2.56	2.61	2.67	2.77	2.79	2.80	2.82	2.84	2.85	2.87	2.89	2.92	2.94	2.97
27	2.29	2.41	2.42	2.43	2.45	2.47	2.49	2.52	2.57	2.63	2.73	2.75	2.76	2.78	2.79	2.81	2.83	2.85	2.88	2.90	2.93
28	2.25	2.37	2.38	2.39	2.41	2.43	2.45	2.48	2.53	2.59	2.69	2.71	2.72	2.74	2.76	2.77	2.79	2.82	2.84	2.86	2.89
29	2.21	2.33	2.34	2.36	2.37	2.39	2.42	2.45	2.49	2.56	2.66	2.67	2.69	2.70	2.72	2.74	2.76	2.78	2.80	2.83	2.86
30	2.18	2.30	2.31	2.32	2.34	2.36	2.38	2.42	2.46	2.52	2.63	2.64	2.66	2.67	2.69	2.71	2.73	2.75	2.77	2.80	2.82
40	1.93	2.06	2.07	2.09	2.10	2.12	2.15	2.18	2.23	2.30	2.40	2.42	2.43	2.45	2.46	2.48	2.50	2.52	2.55	2.57	2.60
50	1.79	1.93	1.94	1.95	1.97	1.99	2.02	2.05	2.10	2.16	2.27	2.29	2.30	2.32	2.33	2.35	2.37	2.39	2.42	2.44	2.47
60	1.69	1.83	1.85	1.86	1.88	1.90	1.93	1.96	2.01	2.08	2.19	2.20	2.22	2.23	2.25	2.27	2.29	2.31	2.33	2.36	2.39
70	1.62	1.77	1.78	1.80	1.81	1.84	1.86	1.90	1.95	2.02	2.13	2.14	2.16	2.17	2.19	2.21	2.23	2.25	2.28	2.30	2.33
80	1.57	1.72	1.73	1.75	1.77	1.79	1.82	1.85	1.90	1.97	2.08	2.10	2.11	2.13	2.15	2.17	2.19	2.21	2.23	2.26	2.29
90	1.52	1.68	1.70	1.71	1.73	1.75	1.78	1.82	1.87	1.94	2.05	2.07	2.08	2.10	2.12	2.13	2.15	2.18	2.20	2.23	2.25
100	1.49	1.65	1.67	1.68	1.70	1.72	1.75	1.79	1.84	1.91	2.02	2.04	2.05	2.07	2.09	2.11	2.13	2.15	2.17	2.20	2.23
110	1.46	1.63	1.64	1.66	1.68	1.70	1.73	1.77	1.82	1.89	2.00	2.02	2.03	2.05	2.07	2.09	2.11	2.13	2.15	2.18	2.21
120	1.43	1.61	1.62	1.64	1.66	1.68	1.71	1.75	1.80	1.87	1.98	2.00	2.01	2.03	2.05	2.07	2.09	2.11	2.13	2.16	2.19
10000	1.00	1.37	1.38	1.40	1.43	1.46	1.49	1.54	1.59	1.67	1.79	1.81	1.82	1.84	1.86	1.88	1.90	1.92	1.95	1.97	2.00

Table B.3 (Continued)

Critical values of the F distribution at 0.001 significance level

	Degrees of freedom, ν_1												
Degrees of freedom, ν_2	1	2	3	4	5	6	7	8	9	10	11	12	13
1	405284	500000	540379	562500	576405	585937	592873	598144	602284	605621	608368	610668	612622
2	998.5	999.0	999.2	999.2	999.3	999.3	999.4	999.4	999.4	999.4	999.4	999.4	999.4
3	167.0	148.5	141.1	137.1	134.6	132.8	131.6	130.6	129.9	129.2	128.7	128.3	128.0
4	74.14	61.25	56.18	53.44	51.71	50.53	49.66	49.00	48.47	48.05	47.70	47.41	47.16
5	47.18	37.12	33.20	31.09	29.75	28.83	28.16	27.65	27.24	26.92	26.65	26.42	26.22
6	35.51	27.00	23.70	21.92	20.80	20.03	19.46	19.03	18.69	18.41	18.18	17.99	17.82
7	29.25	21.69	18.77	17.20	16.21	15.52	15.02	14.63	14.33	14.08	13.88	13.71	13.56
8	25.41	18.49	15.83	14.39	13.48	12.86	12.40	12.05	11.77	11.54	11.35	11.19	11.06
9	22.86	16.39	13.90	12.56	11.71	11.13	10.70	10.37	10.11	9.89	9.72	9.57	9.44
10	21.04	14.91	12.55	11.28	10.48	9.93	9.52	9.20	8.96	8.75	8.59	8.45	8.32
11	19.69	13.81	11.56	10.35	9.58	9.05	8.66	8.35	8.12	7.92	7.76	7.63	7.51
12	18.64	12.97	10.80	9.63	8.89	8.38	8.00	7.71	7.48	7.29	7.14	7.00	6.89
13	17.82	12.31	10.21	9.07	8.35	7.86	7.49	7.21	6.98	6.80	6.65	6.52	6.41
14	17.14	11.78	9.73	8.62	7.92	7.44	7.08	6.80	6.58	6.40	6.26	6.13	6.02
15	16.59	11.34	9.34	8.25	7.57	7.09	6.74	6.47	6.26	6.08	5.94	5.81	5.71
16	16.12	10.97	9.01	7.94	7.27	6.80	6.46	6.19	5.98	5.81	5.67	5.55	5.44
17	15.72	10.66	8.73	7.68	7.02	6.56	6.22	5.96	5.75	5.58	5.44	5.32	5.22

18	15.38	10.39	8.49	7.46	6.81	6.35	6.02	5.76	5.56	5.39	5.25	5.13	5.03
19	15.08	10.16	8.28	7.27	6.62	6.18	5.85	5.59	5.39	5.22	5.08	4.97	4.87
20	14.82	9.95	8.10	7.10	6.46	6.02	5.69	5.44	5.24	5.08	4.94	4.82	4.72
21	14.59	9.77	7.94	6.95	6.32	5.88	5.56	5.31	5.11	4.95	4.81	4.70	4.60
22	14.38	9.61	7.80	6.81	6.19	5.76	5.44	5.19	4.99	4.83	4.70	4.58	4.49
23	14.20	9.47	7.67	6.70	6.08	5.65	5.33	5.09	4.89	4.73	4.60	4.48	4.39
24	14.03	9.34	7.55	6.59	5.98	5.55	5.23	4.99	4.80	4.64	4.51	4.39	4.30
25	13.88	9.22	7.45	6.49	5.89	5.46	5.15	4.91	4.71	4.56	4.42	4.31	4.22
26	13.74	9.12	7.36	6.41	5.80	5.38	5.07	4.83	4.64	4.48	4.35	4.24	4.14
27	13.61	9.02	7.27	6.33	5.73	5.31	5.00	4.76	4.57	4.41	4.28	4.17	4.08
28	13.50	8.93	7.19	6.25	5.66	5.24	4.93	4.69	4.50	4.35	4.22	4.11	4.01
29	13.39	8.85	7.12	6.19	5.59	5.13	4.87	4.64	4.45	4.29	4.16	4.05	3.96
30	13.29	8.77	7.05	6.12	5.53	5.12	4.82	4.58	4.39	4.24	4.11	4.00	3.91
40	12.61	8.25	6.59	5.70	5.13	4.73	4.44	4.21	4.02	3.87	3.75	3.64	3.55
50	12.22	7.96	6.34	5.46	4.90	4.51	4.22	4.00	3.82	3.67	3.55	3.44	3.35
60	11.97	7.77	6.17	5.31	4.76	4.37	4.09	3.86	3.69	3.54	3.42	3.32	3.23
70	11.80	7.64	6.06	5.20	4.66	4.23	3.99	3.77	3.60	3.45	3.33	3.23	3.14
80	11.67	7.54	5.97	5.12	4.58	4.21	3.92	3.70	3.53	3.39	3.27	3.16	3.07
90	11.57	7.47	5.91	5.06	4.53	4.15	3.87	3.65	3.48	3.34	3.22	3.11	3.02
100	11.50	7.41	5.86	5.02	4.48	4.11	3.83	3.61	3.44	3.30	3.18	3.07	2.99
110	11.43	7.36	5.82	4.98	4.45	4.07	3.79	3.58	3.41	3.26	3.14	3.04	2.95
120	11.38	7.32	5.78	4.95	4.42	4.04	3.77	3.55	3.38	3.24	3.12	3.02	2.93
10000	10.83	6.91	5.43	4.62	4.11	3.75	3.48	3.27	3.10	2.96	2.85	2.75	2.66

Table B.3 (Continued)

Critical values of the F distribution at 0.001 significance level

	Degrees of freedom, ν_1													
Degrees of freedom, ν_2	14	15	16	17	18	19	20	21	22	23	24	25	26	27
1	614303	615764	617045	618178	619188	620092	620908	621646	622319	622933	623497	624017	624497	624941
2	999.4	999.4	999.4	999.4	999.4	999.4	999.4	999.5	999.5	999.5	999.5	999.5	999.5	999.5
3	127.6	127.4	127.1	126.9	126.7	126.6	126.4	126.3	126.2	126.0	125.9	125.8	125.7	125.7
4	46.95	46.76	46.60	46.45	46.32	46.21	46.10	46.00	45.92	45.84	45.77	45.70	45.64	45.58
5	26.06	25.91	25.78	25.67	25.57	25.48	25.39	25.32	25.25	25.19	25.13	25.08	25.03	24.99
6	17.68	17.56	17.45	17.35	17.27	17.19	17.12	17.06	17.00	16.95	16.90	16.85	16.81	16.77
7	13.43	13.32	13.23	13.14	13.06	12.99	12.93	12.87	12.82	12.78	12.73	12.69	12.65	12.62
8	10.94	10.84	10.75	10.67	10.60	10.54	10.48	10.43	10.38	10.34	10.30	10.26	10.22	10.19
9	9.33	9.24	9.15	9.08	9.01	8.95	8.90	8.85	8.80	8.76	8.72	8.69	8.66	8.63
10	8.22	8.13	8.05	7.98	7.91	7.86	7.80	7.76	7.71	7.67	7.64	7.60	7.57	7.54
11	7.41	7.32	7.24	7.17	7.11	7.06	7.01	6.96	6.92	6.88	6.85	6.81	6.78	6.76
12	6.79	6.71	6.63	6.57	6.51	6.45	6.40	6.36	6.32	6.28	6.25	6.22	6.19	6.16
13	6.31	6.23	6.16	6.09	6.03	5.98	5.93	5.89	5.85	5.81	5.78	5.75	5.72	5.70
14	5.93	5.85	5.78	5.71	5.66	5.60	5.56	5.51	5.48	5.44	5.41	5.38	5.35	5.32
15	5.62	5.54	5.46	5.40	5.35	5.29	5.25	5.21	5.17	5.13	5.10	5.07	5.04	5.02
16	5.35	5.27	5.20	5.14	5.09	5.04	4.99	4.95	4.91	4.88	4.85	4.82	4.79	4.76
17	5.13	5.05	4.99	4.92	4.87	4.82	4.78	4.73	4.70	4.66	4.63	4.60	4.57	4.55

18	4.94	4.87	4.80	4.74	4.68	4.63	4.59	4.55	4.51	4.48	4.45	4.42	4.39	4.37
19	4.78	4.70	4.64	4.58	4.52	4.47	4.43	4.39	4.35	4.32	4.29	4.26	4.23	4.21
20	4.64	4.56	4.49	4.44	4.38	4.33	4.29	4.25	4.21	4.18	4.15	4.12	4.09	4.07
21	4.51	4.44	4.37	4.31	4.26	4.21	4.17	4.13	4.09	4.06	4.03	4.00	3.97	3.95
22	4.40	4.33	4.26	4.20	4.15	4.10	4.06	4.02	3.98	3.95	3.92	3.89	3.86	3.84
23	4.30	4.23	4.16	4.10	4.05	4.00	3.96	3.92	3.89	3.85	3.82	3.79	3.77	3.74
24	4.21	4.14	4.07	4.02	3.96	3.92	3.87	3.83	3.80	3.77	3.74	3.71	3.68	3.66
25	4.13	4.06	3.99	3.94	3.88	3.84	3.79	3.76	3.72	3.69	3.66	3.63	3.60	3.58
26	4.06	3.99	3.92	3.86	3.81	3.77	3.72	3.68	3.65	3.62	3.59	3.56	3.53	3.51
27	3.99	3.92	3.86	3.80	3.75	3.70	3.66	3.62	3.58	3.55	3.52	3.49	3.47	3.44
28	3.93	3.86	3.80	3.74	3.69	3.64	3.60	3.56	3.52	3.49	3.46	3.43	3.41	3.38
29	3.88	3.80	3.74	3.68	3.63	3.59	3.54	3.50	3.47	3.44	3.41	3.38	3.35	3.33
30	3.82	3.75	3.69	3.63	3.58	3.53	3.49	3.45	3.42	3.39	3.36	3.33	3.30	3.28
40	3.47	3.40	3.34	3.28	3.23	3.19	3.14	3.11	3.07	3.04	3.01	2.98	2.96	2.93
50	3.27	3.20	3.14	3.09	3.04	2.99	2.95	2.91	2.88	2.85	2.82	2.79	2.76	2.74
60	3.15	3.08	3.02	2.96	2.91	2.87	2.83	2.79	2.75	2.72	2.69	2.67	2.64	2.62
70	3.06	2.99	2.93	2.88	2.83	2.78	2.74	2.70	2.67	2.54	2.61	2.58	2.56	2.53
80	3.00	2.93	2.87	2.81	2.76	2.72	2.68	2.64	2.61	2.57	2.54	2.52	2.49	2.47
90	2.95	2.88	2.82	2.76	2.71	2.67	2.63	2.59	2.56	2.53	2.50	2.47	2.44	2.42
100	2.91	2.84	2.78	2.73	2.68	2.63	2.59	2.55	2.52	2.49	2.46	2.43	2.41	2.38
110	2.88	2.81	2.75	2.59	2.65	2.60	2.56	2.52	2.49	2.46	2.43	2.40	2.37	2.35
120	2.85	2.78	2.72	2.67	2.62	2.58	2.53	2.50	2.46	2.43	2.40	2.37	2.35	2.33
10000	2.58	2.52	2.46	2.40	2.35	2.31	2.27	2.23	2.20	2.17	2.14	2.11	2.08	2.06

Table B.3 (Continued)

Critical values of the F distribution at 0.001 significance level

Degrees of freedom, ν_2	Degrees of freedom, ν_1												
	28	29	30	40	50	60	70	80	90	100	110	120	10000
1	625354	625739	626099	628712	630285	631337	632089	632653	633093	633444	633732	633972	636588
2	999.5	999.5	999.5	999.5	999.5	999.5	999.5	999.5	999.5	999.5	999.5	999.5	999.5
3	125.6	125.5	125.4	125.0	124.7	124.5	124.3	124.2	124.1	124.1	124.0	124.0	123.5
4	45.53	45.48	45.43	45.09	44.88	44.75	44.65	44.57	44.52	44.47	44.43	44.40	44.06
5	24.94	24.91	24.87	24.60	24.44	24.33	24.26	24.20	24.15	24.12	24.09	24.06	23.79
6	16.74	16.70	16.67	16.44	16.31	16.21	16.15	16.10	16.06	16.03	16.00	15.98	15.75
7	12.59	12.56	12.53	12.33	12.20	12.12	12.06	12.01	11.98	11.95	11.93	11.91	11.70
8	10.16	10.13	10.11	9.92	9.80	9.73	9.67	9.63	9.60	9.57	9.55	9.53	9.34
9	8.60	8.57	8.55	8.37	8.26	8.19	8.13	8.09	8.06	8.04	8.02	8.00	7.82
10	7.52	7.49	7.47	7.30	7.19	7.12	7.07	7.03	7.00	6.98	6.96	6.94	6.76
11	6.73	6.71	6.68	6.52	6.42	6.35	6.30	6.26	6.23	6.21	6.19	6.18	6.00
12	6.14	6.11	6.09	5.93	5.83	5.76	5.71	5.68	5.65	5.63	5.61	5.59	5.42
13	5.67	5.65	5.63	5.47	5.37	5.30	5.26	5.22	5.19	5.17	5.15	5.14	4.97
14	5.30	5.28	5.25	5.10	5.00	4.94	4.89	4.86	4.83	4.81	4.79	4.77	4.61
15	4.99	4.97	4.95	4.80	4.70	4.64	4.59	4.56	4.53	4.51	4.49	4.47	4.31
16	4.74	4.72	4.70	4.54	4.45	4.39	4.34	4.31	4.28	4.26	4.24	4.23	4.06
17	4.53	4.50	4.48	4.33	4.24	4.18	4.13	4.10	4.07	4.05	4.03	4.02	3.85

18	4.34	4.32	4.30	4.15	4.06	4.00	3.95	3.92	3.89	3.87	3.85	3.84	3.67
19	4.18	4.16	4.14	3.99	3.90	3.84	3.79	3.76	3.73	3.71	3.69	3.68	3.52
20	4.05	4.03	4.00	3.86	3.77	3.70	3.66	3.62	3.60	3.58	3.56	3.54	3.38
21	3.93	3.90	3.88	3.74	3.64	3.58	3.54	3.50	3.48	3.46	3.44	3.42	3.26
22	3.82	3.80	3.78	3.63	3.54	3.48	3.43	3.40	3.37	3.35	3.33	3.32	3.15
23	3.72	3.70	3.68	3.53	3.44	3.38	3.34	3.30	3.28	3.25	3.24	3.22	3.06
24	3.63	3.61	3.59	3.45	3.36	3.29	3.25	3.22	3.19	3.17	3.15	3.14	2.97
25	3.56	3.54	3.52	3.37	3.28	3.22	3.17	3.14	3.11	3.09	3.07	3.06	2.89
26	3.49	3.46	3.44	3.30	3.21	3.15	3.10	3.07	3.04	3.02	3.00	2.99	2.82
27	3.42	3.40	3.38	3.23	3.14	3.08	3.04	3.00	2.98	2.96	2.94	2.92	2.75
28	3.36	3.34	3.32	3.18	3.09	3.03	2.98	2.94	2.92	2.90	2.88	2.86	2.70
29	3.31	3.29	3.27	3.12	3.03	2.97	2.92	2.89	2.86	2.84	2.82	2.81	2.64
30	3.26	3.24	3.22	3.07	2.98	2.92	2.87	2.84	2.81	2.79	2.77	2.76	2.59
40	2.91	2.89	2.87	2.73	2.64	2.57	2.53	2.49	2.47	2.44	2.43	2.41	2.23
50	2.72	2.70	2.68	2.53	2.44	2.33	2.33	2.30	2.27	2.25	2.23	2.21	2.03
60	2.60	2.57	2.55	2.41	2.32	2.25	2.21	2.17	2.14	2.12	2.10	2.08	1.89
70	2.51	2.49	2.47	2.32	2.23	2.15	2.12	2.08	2.05	2.03	2.01	1.99	1.80
80	2.45	2.43	2.41	2.26	2.16	2.10	2.05	2.01	1.98	1.96	1.94	1.92	1.72
90	2.40	2.38	2.36	2.21	2.11	2.05	2.00	1.96	1.93	1.91	1.89	1.87	1.66
100	2.36	2.34	2.32	2.17	2.08	2.01	1.96	1.92	1.89	1.87	1.85	1.83	1.52
110	2.33	2.31	2.29	2.14	2.04	1.98	1.93	1.89	1.86	1.83	1.81	1.80	1.58
120	2.30	2.28	2.26	2.11	2.02	1.55	1.90	1.86	1.83	1.81	1.78	1.77	1.55
10000	2.04	2.01	1.99	1.84	1.74	1.66	1.61	1.56	1.53	1.50	1.47	1.45	1.00

Table B.4 Critical values of the U-distribution.

Critical values of the U-distribution at 0.05 significance level (one-tailed) or 0.1 significance level (two-tailed)

n_x \ n_y	1	2	3	4	5	6	7	8	9	10	11	12	13	14	15	16	17	18	19	20
1																			0	0
2					0	0	0	1	1	1	1	2	2	2	3	3	3	4	4	4
3			0	0	1	2	2	3	3	4	5	5	6	7	7	8	9	9	10	11
4			0	1	2	3	4	5	6	7	8	9	10	11	12	14	15	16	17	18
5		0	1	2	4	5	6	8	9	11	12	13	15	16	18	19	20	22	23	25
6		0	2	3	5	7	8	10	12	14	16	17	19	21	23	25	26	28	30	32
7		0	2	4	6	8	11	13	15	17	19	21	24	26	28	30	33	35	37	39
8		1	3	5	8	10	13	15	18	20	23	26	28	31	33	36	39	41	44	47
9		1	3	6	9	12	15	18	21	24	27	30	33	36	39	42	45	48	51	54
10		1	4	7	11	14	17	20	24	27	31	34	37	41	44	48	51	55	58	62
11		1	5	8	12	16	19	23	27	31	34	38	42	46	50	54	57	61	65	69
12		2	5	9	13	17	21	26	30	34	38	42	47	51	55	60	64	68	72	77

13		2	6	10	15	19	24	28	33	37	42	47	51	56	61	65	70	75	80	84
14		2	7	11	16	21	26	31	36	41	46	51	56	61	66	71	77	82	87	92
15		3	7	12	18	23	28	33	39	44	50	55	61	66	72	77	83	88	94	100
16		3	8	14	19	25	30	36	42	48	54	60	65	71	77	83	89	95	101	107
17		3	9	15	20	26	33	39	45	51	57	64	70	77	83	89	96	102	109	115
18		4	9	16	22	28	35	41	48	55	61	68	75	82	88	95	102	109	116	123
19	0	4	10	17	23	30	37	44	51	58	65	72	80	87	94	101	109	116	123	130
20	0	4	11	18	25	32	39	47	54	62	69	77	84	92	100	107	115	123	130	138

Reject H_0 if the calculated value of U is less than or equal to the critical value in the table

Table B.4 (Continued)

Critical values of the U-distribution at 0.025 significance level (one-tailed) or 0.05 significance level (two-tailed)

	n_y																			
n_x	1	2	3	4	5	6	7	8	9	10	11	12	13	14	15	16	17	18	19	20
1																				
2								0	0	0	0	1	1	1	1	1	2	2	2	2
3					0	1	1	2	2	3	3	4	4	5	5	6	6	7	7	8
4				0	1	2	3	4	4	5	6	7	8	9	10	11	11	12	13	13
5			0	1	2	3	5	6	7	8	9	11	12	13	14	15	17	18	19	20
6			1	2	3	5	6	8	10	11	13	14	16	17	19	21	22	24	25	27
7			1	3	5	6	8	10	12	14	16	18	20	22	24	26	28	30	32	34
8		0	2	4	6	8	10	13	15	17	19	22	24	26	29	31	34	36	38	41
9		0	2	4	7	10	12	15	17	20	23	26	28	31	34	37	39	42	45	48
10		0	3	5	8	11	14	17	20	23	26	29	33	36	39	42	45	48	52	55

11	0	3	6	9	13	16	19	23	26	30	33	37	40	44	47	51	55	58	62
12	1	4	7	11	14	18	22	25	29	33	37	41	45	49	53	57	61	65	69
13	1	4	8	12	16	20	24	28	33	37	41	45	50	54	59	63	67	72	76
14	1	5	9	13	17	22	26	31	36	40	45	50	55	59	64	69	74	78	83
15	1	5	10	14	19	24	29	34	39	44	49	54	59	64	70	75	80	85	90
16	1	6	11	15	21	26	31	37	42	47	53	59	64	70	75	81	86	92	98
17	2	6	11	17	22	28	34	39	45	51	57	63	67	75	81	87	93	99	105
18	2	7	12	18	24	30	36	42	48	55	61	67	74	80	86	93	99	106	112
19	2	7	13	19	25	32	38	45	52	58	65	72	78	85	92	99	106	113	119
20	2	8	13	20	27	34	41	48	55	62	69	76	83	90	98	105	112	119	127

Table B.5 Critical values of the χ^2 distribution.

	Significance level				
ν	0.1	0.05	0.01	0.005	0.001
1	2.71	3.84	6.63	7.88	10.83
2	4.61	5.99	9.21	10.60	13.82
3	6.25	7.81	11.34	12.84	16.27
4	7.78	9.49	13.28	14.86	18.47
5	9.24	11.07	15.09	16.75	20.52
6	10.64	12.59	16.81	18.55	22.46
7	12.02	14.07	18.48	20.28	24.32
8	13.36	15.51	20.09	21.95	26.12
9	14.68	16.92	21.67	23.59	27.88
10	15.99	18.31	23.21	25.19	29.59
11	17.28	19.68	24.72	26.76	31.26
12	18.55	21.03	26.22	28.30	32.91
13	19.81	22.36	27.69	29.82	34.53
14	21.06	23.68	29.14	31.32	36.12
15	22.31	25.00	30.58	32.80	37.70
16	23.54	26.30	32.00	34.27	39.25
17	24.77	27.59	33.41	35.72	40.79
18	25.99	28.87	34.81	37.16	42.31
19	27.20	30.14	36.19	38.58	43.82
20	28.41	31.41	37.57	40.00	45.31
21	29.62	32.67	38.93	41.40	46.80
22	30.81	33.92	40.29	42.80	48.27
23	32.01	35.17	41.64	44.18	49.73
24	33.20	36.42	42.98	45.56	51.18
25	34.38	37.65	44.31	46.93	52.62
26	35.56	38.89	45.64	48.29	54.05
27	36.74	40.11	46.96	49.64	55.48
28	37.92	41.34	48.28	50.99	56.89
29	39.09	42.56	49.59	52.34	58.30

Table B.5 (Continued)

ν	Significance level				
	0.1	0.05	0.01	0.005	0.001
30	40.26	43.77	50.89	53.67	59.70
40	51.81	55.76	63.69	66.77	73.40
50	63.17	67.50	76.15	79.49	86.66
60	74.40	79.08	88.38	91.95	99.61
70	85.53	90.53	100.43	104.21	112.32
80	96.58	101.88	112.33	116.32	124.84
90	107.57	113.15	124.12	128.30	137.21
100	118.50	124.34	135.81	140.17	149.45

Reject H_0 if the calculated level of χ^2 is greater than the critical value at the chosen significance level.

References

Atkinson, A. C. (1982) Regression diagnostics, transformations and constructed variables. *Journal of the Royal Statistical Society: Series B (Methodological)*, 44(1), 1–36.

Chatfield, C. (1996) *The Analysis of Time Series: An Introduction*. Chapman and Hall, London.

Cliff, A. D. and Ord, J. K. (1981) *Spatial Processes, Models and Applications*. Pion, London.

Crawley, M. J. (2005) *Statistics: An Introduction Using R*. John Wiley & Sons, Ltd, Chichester.

Crawley, M. J. (2012) *The R Book*. John Wiley & Sons, Ltd, Chichester.

Cressie, N. A. C. (1993) *Statistics for Spatial Data*. John Wiley & Sons, Inc., New York.

Diamond, J. M. and Mayr, E. (1976) Species–area relation for birds of the Solomon Archipelago. *Proceedings of the National Academy of Sciences of the United States of America*, 73(1), 262–266.

Diggle, P. (1990) *Time Series: A Biostatistical Introduction*. Clarendon Press, Oxford.

Diggle, P. (2002) *Statistical Analysis of Spatial Point Patterns*. Hodder, London.

Draper, N. R. and Smith, H. (1981) *Applied Regression Analysis*, 2nd edn. John Wiley & Sons, Ltd, Chichester.

Evenden, G. I. (2003) Cartographic projection procedures for the UNIX environment – A user's manual. *United States Geological Survey Open File Report* 90-284.

Gwiazda, J., Ong, E., Held, R. and Thorn, F. (2000) Myopia and ambient night-time lighting. *Nature*, 404, 144.

Statistical Analysis of Geographical Data: An Introduction, First Edition.
Simon J. Dadson.
© 2017 John Wiley & Sons Ltd. Published 2017 by John Wiley & Sons Ltd.

Mark, D. M. and Church, M. A. (1977) On the misuse of regression in Earth science. *Mathematical Geology*, 9(1), 63–75.

Montgomery, D. C. (2008) *Design and Analysis of Experiments*, 7th edn. John Wiley & Sons, Ltd, Chichester.

NIMA (1997) World Geodetic System 1984: Its definition and relationships with local geodetic systems. *National Imagery and Mapping Agency Technical Report* TR8350.2.

Quinn, G. E., Shin, C. H., Maguire, M. G. and Stone, R. A. (1999) Myopia and ambient lighting at night. *Nature*, 399, 113.

R Core Team (2015) *R: A language and environment for statistical computing*. R Foundation for Statistical Computing, Vienna. http://www.r-project.org/.

Roscoe, J. T. and Byars, J. A. (1971) An investigation of the restraints with respect to sample size commonly imposed on the use of the chi-square statistic. *Journal of the American Statistical Association*, 66(336), 755–759

Selvey-Clinton, P. (2006) The contamination and distribution of heavy metals along the River Twymyn, Dylife, Wales. Dissertation, University of Oxford.

Snyder, J. P. (1987) Map Projections: A working manual. *United States Geological Survey Professional Paper* 1395.

Squire, P. (1988) Why the 1936 Literary Digest poll failed. *The Public Opinion Quarterly*, 52(1), 125–133.

Townsend, C. R., Hildrew, A. G. and Francis, J. (1983) Community structure in some southern English streams: the influence of physicochemical factors. *Freshwater Biology*, 13, 521–544.

Venables, W. N. and Ripley, B. D. (2003) *Modern Applied Statistics with S*. Springer, New York.

Zadnik, K., Jones, L. A., Irvin, B. C. *et al.* (2000) Vision: Myopia and ambient night-time lighting. *Nature*, 404, 143–144.

Zar, J. H. (1972) Significance testing of the Spearman rank correlation coefficient. *Journal of the American Statistical Association*, 67(339), 578–580.

Index

a

accuracy
  data 6–7, 8
  vs. precision 6–7, 8
analysis of residuals, linear
    regression 128 130, 131
analysis of variance (ANOVA)
  89–108
  applications 106–107
  assumptions 99–101
  Barlett's test 101
  diagnostics 99–101
  exercises 107–108
  F-ratio 96–97
  Kruskal–Wallis H-test
    105–106
  linear regression 133–135
  mean squared values (MS) 96
  multiple comparison tests
    101–105
  non-parametric methods
    105–106
  nuisance factors 106–107
  one-way analysis of variance
    90–99
  randomized experimental
    design 90–91
  R statistical software
    97–99
  two-way analysis of
    variance 107
assumptions
  analysis of variance (ANOVA)
    99–101
  Chi-squared test ($\chi2$) 82
  F-test 74
  linear regression 130–131,
    140–141
  one-sample Z-test 56
  parametric tests vs.
    non-parametric tests 75
  probability 42
  stationarity 180
  time series analysis 180
autocorrelation, time series
  analysis 189–190

b

Barlett's test
  see also F-test
  analysis of variance
    (ANOVA) 101
basic statistics, R statistical
  software 197

Statistical Analysis of Geographical Data: An Introduction, First Edition.
Simon J. Dadson.
© 2017 John Wiley & Sons Ltd. Published 2017 by John Wiley & Sons Ltd.

bias
  research design 14–15
  sampling frame 14–15
  sampling methods 43
  stratified sampling 17
bimodal frequency distribution
  18–19

c

causality and correlation
  116–117
central limit theorem 43–46
  sampling distributions
    43–46
central tendency of datasets
  24–28
  mean 24–28
  median 25–28
  mode 25
  worked example 26–28
Chebyshev's theorem 30
Chi-squared test ($\chi2$) 81–88
  assumptions 82
  clusters 161–162
  critical values of the $\chi2$
    distribution (statistical
    table) 238–239
  Excel 84, 86
  one sample 81–84
  R statistical software 84, 86
  two samples 84–88
clusters 159–162
  Chi-squared test ($\chi2$)
    161–162
  nearest neighbour
    statistics 162
  quadrat test 159–162
  spatial statistics 159–162
coefficient of determination, linear
  regression 135–137
coefficient of variation
  exercise 35

summarizing data numerically
  30–33
  worked example 31–33
comparing datasets,
  exercises 79
comparing distributions
  Chi-squared test ($\chi2$) 81–88
  exercises 87–88
comparing means, t-test 61–70
comparing proportions 63–64
comparing samples 64–75
  independent samples 64
comparing variances,
  F-test 74–75
confidence intervals 49–54
  exercises 54
  linear regression
    137–140, 141
  normal distribution 49–50
  for a proportion 53–54
  sample size 53
  t-distribution 50–53
contour maps, spatial statistics
  162–163
coordinate systems, spatial data
  types 5–6
correlation 109–119
  autocorrelation, time series
    analysis 189–190
  correlograms 190, 191
  exercises 117–119
correlation analysis 109–110
correlation and causality
  116–117
  hidden variables 116–117
  sample size 117
  Type I error (false
    positive) 117
correlation coefficient
  Pearson's product-moment
    correlation coefficient
    110–112

significance tests of correlation
coefficient   112–114
Spearman's rank correlation
coefficient   114–116

d

data
exercises   12
plotting data, R statistical
software   197–199
reading data, R statistical
software   202–204
reporting data and
uncertainties   7–9
writing data, R statistical
software   202–204
datasets
central tendency of datasets
24–28
comparison exercises   79
data types   1–6
interval data   3–4
nominal data   3–4
ordinal data   3–4
raster (or gridded) data
5–6, 146
ratio data   3–4
spatial data types   5–6, 145–146
vector data   5
decomposing time series, time
series analysis   179–180
degrees of freedom, variance   30
density estimation, spatial data
158–159
diagnostics, analysis of variance
(ANOVA)   99–101
dispersion
inter-quartile range   28
summarizing data numerically
28–29
distributions, comparison
exercises   87–88

e

error, measurement see
measurement error
examples, worked see worked
examples
Excel
see also R statistical software
Chi-squared test ($\chi 2$)   84, 86
frequency distributions   20
Mann–Whitney U-test
77–78
scatter plots   124
standard deviation   31–33
t-test   67
variance   31–33
exercises
analysis of variance (ANOVA)
107–108
coefficient of variation   35
comparing datasets   79
comparing distributions
87–88
confidence intervals   54
correlation   117–119
data   12
linear regression   142–144
mean   33–35
probability   47
sampling distributions   47
scientific notation (standard
form)   12
spatial statistics   171
standard deviation   33–35
standard error of the mean   47
time series analysis   190–192

f

false negative see Type II error
false positive see Type I error
F-distribution, critical values of
the F-distribution (statistical
table)   212–233

F-ratio, analysis of variance
  (ANOVA)   96–97
frequency distributions   17–21
  bimodal   18–19
  Excel   20
  leptokurtic   18–19
  multimodal   18–19
  negative skew   18–19
  platykurtic   18–19
  positive skew   18–19
  R statistical software   21
  worked example   19–21
F-test   97, 133
  see also Barlett's test
  assumptions   74
  comparing variances   74–75
  critical values of the
    F-distribution (statistical
    table)   212–233

g
Global Positioning System (GPS),
  map projections   150
graphical summaries   17–24
  frequency distributions   17–21
  histograms   17–21
  scatter plots   22–24
  time series plots   21–22
gridded (raster) data   5–6, 146

h
hidden variables, correlation and
  causality   116–117
histograms   17–21
hypothesis testing   55–61
  general procedure   61
  linear regression   137–140
  non- parametric hypothesis
    testing   75–79
  null hypothesis   56–61
  one-sample *Z*-test   56–61

parametric tests vs.
  non-parametric tests   75
p-values   60–61
Type I error (false positive)
  57–58
Type II error (false negative)
  57–58
Z-test   56–61

i
interpolation, spatial statistics
  162–163
inter-quartile range,
  dispersion   28
interval data, data type   3–4
irregular fluctuations, time series
  analysis   179–180

j
join counts
  R statistical software
    167–170
  spatial relationships
    164–170

k
Kruskal–Wallis *H*-test, analysis
  of variance (ANOVA)
    105–106
kurtosis
  leptokurtic frequency
    distribution   18–19
  platykurtic frequency
    distribution   18–19
  summarizing data
    numerically   33

l
Lambert conformal conic
  projection, vs. Mercator
  projection   153–157

latitude/longitude 146–149
least-squares linear regression
    121–122
'least-squares' procedure, linear
    regression 124–128
leptokurtic frequency
    distribution 18–19
linear regression 121–144
  analysis of residuals
    128–130, 131
  analysis of variance (ANOVA)
    133–135
  assumptions 130–131,
    140–141
  caveats 130–131
  coefficient of determination
    135–137
  confidence intervals
    137–140, 141
  exercises 142–144
  hypothesis testing 137–140
  least-squares linear regression
    121–122
  'least-squares' procedure
    124–128
  line of best fit 124–128
  reduced major axis regression
    140–142
  residuals analysis 128–130, 131
  scatter plots 122–124
  significance 131
  standard error of the
    intercept 137
  standard error of the slope 137
  tests on the regression
    parameters 138–139
line of best fit, linear regression
    124–128
London Mayoral Election of
    2012: 164–170
longitude/latitude 146–149

m
Mann–Whitney *U*-test
    see also Wilcoxon rank
      sum test
  critical values of the
    U-distribution (statistical
    table) 234–237
  Excel 77–78
map projections
  Global Positioning System
    (GPS) 150
  Lambert conformal conic
    projection 153–157
  Mercator projection 150,
    151–156
  polar stereographic projection
    152–153
  R statistical software
    154–157
  spatial data 149 156
  transverse Mercator projection
    150, 151–152, 154, 155
  Universal Polar Stereographic
    (UPS) projection
    151, 153
  UTM (Universal Transverse
    Mercator) system 151,
    152–153
  World Geodetic System 1984
    (WGS84) reference
    ellipsoid 150
mean
  central tendency of datasets
    24–28
  comparing 61–70
  exercises 33–35
  one-sample *t*-test 61–70
mean centre, spatial data 157
mean squared values (MS),
    analysis of variance
    (ANOVA) 96

measurement error 6–10
  reporting data and
    uncertainties 7–9
  rounding errors 9–10
  significant figures 9–10
median, central tendency of
  datasets 25–28
Mercator projection 151–156
  vs. Lambert conformal conic
    projection 153–157
  transverse Mercator projection
    150, 151–152, 154 155
  UTM (Universal Transverse
    Mercator) system 151,
    152–153
mode, central tendency of
  datasets 25
moving average, time series
  analysis 181–186, 187
MS *see* mean squared values
multimodal frequency
  distribution 18–19
multiple comparison tests,
  analysis of variance
  (ANOVA) 101–105
multiple figures, R statistical
  software 199–202

n
nearest neighbour statistics,
  clusters 162
negative skew, frequency
  distributions 18–19
nominal data, data type 3–4
non- parametric hypothesis
  testing 75–79
  Mann–Whitney *U*-test 76–79
  parametric tests vs.
    non-parametric tests 75
  Wilcoxon rank sum test
    78–79

non-parametric methods
  analysis of variance (ANOVA)
    105–106
  Kruskal–Wallis *H*-test
    105–106
  Mann–Whitney *U*-test
    76–79
normal distribution
  cf. *t*-distribution 51–52
  confidence intervals 49–50
  probability 38–42
  probability density function
    (PDF) 38–39
  properties 38–39
  z-score 39–42
nuisance factors
  analysis of variance (ANOVA)
    106–107
  random sampling 106–107
null hypothesis, hypothesis
  testing 56–61

o
one-sample *t*-test 61–70
one-sample *Z*-test
  assumptions 56
  hypothesis testing 56–61
one-way analysis of variance
  90–99
ordinal data, data type 3–4

p
paired samples (paired *t*-test)
  71–73
parametric tests vs.
    non-parametric tests 75
PDF *see* probability density
    function
Pearson's product-moment
    correlation coefficient
    110–112

platykurtic frequency
distribution   18–19
plotting data, R statistical
software   197–199
polar stereographic projection
151, 152–153
population density,
calculations   12
positive skew, frequency
distributions   18–19
precision
vs. accuracy   6–7, 8
data   6–7, 8
probability   37–47
assumptions   42
exercises   47
normal distribution   38–42
random variables   37–38
standard deviation   39–45
z-score   39–42
probability density function
(PDF), normal distribution
38–39
projections, map *see* map
projections
proportion, confidence intervals
53–54
proportions, comparing   63–64
p-values, hypothesis testing
60–61

q
quadrat test, clusters   159–162

r
randomized experimental design,
analysis of variance
(ANOVA)   90–91
random numbers   15
random sampling   15–16
nuisance factors   106–107

raster (gridded) data   5–6, 146
ratio data, data type   3–4
reading data, R statistical
software   202–204
reduced major axis regression,
linear regression   140–142
regression, linear *see* linear
regression
reporting data and uncertainties
7–9
research design
bias   14–15
sampling methods   13–15
residuals analysis, linear
regression   128–130, 131
role of statistics   1–2
rounding errors, measurement
error   9–10
R statistical software   193–204
see also Excel
analysis of variance (ANOVA)
97–99
basic statistics   197
Chi-squared test ($\chi 2$)   84, 86
frequency distributions   21
join counts   167–170
map projections   154–157
multiple figures   199–202
obtaining   193
plotting data   197–199
reading data   202–204
scatter plots   124
seasonal variation   188–189
simple calculations   193–196
spatial data   147–149
time series analysis   178–179
t-test   67–68
vectors   196–197
Wilcoxon rank sum test
78–79
writing data   202–204

S

samples, comparing 64–75
sample size
  confidence intervals 53
  correlation and causality 117
sampling distributions
  central limit theorem 43–46
  exercises 47
sampling frame 14
sampling methods 13–17
  bias 43
  random sampling 15–16
  research design 13–15
  stratified sampling 17
  systematic sampling 16
scatter plots
  Excel 124
  graphical summaries 22–24
  linear regression 122–124
  R statistical software 124
scientific notation (standard
    form) 10–12
  calculations 11–12
  exercise 12
  worked example 12
seasonality, time series analysis
    179–180
seasonal variation
  R statistical software
    188–189
  time series analysis 187–189
significance, linear
    regression 131
significance tests of correlation
    coefficient 112–114
significant figures, measurement
    error 9–10
simple calculations, R statistical
    software 193–196
skewness
  negative skew 18–19
  positive skew 18–19

summarizing data
  numerically 33
software *see* Excel; R statistical
    software
spacial autocorrelation 163–164
spatial data 145–158
  density estimation 158–159
  latitude/longitude 146–149
  map projections 149–156
  mean centre 157
  R statistical software 147–149
  structures 146–149
  summarizing spatial data
    157–159
  weighted mean centre
    157–158
spatial data types 5–6, 145–146
  coordinate systems 5–6
  raster (or gridded) data
    5–6, 146
  vector data 5
spatial relationships 163–170
  join counts 164–170
  spacial autocorrelation
    163–164
  spatial statistics 163–170
spatial statistics 145–170
  clusters 159–162
  contour maps 162–163
  exercises 171
  interpolation 162–163
  spatial data 145–158
  spatial relationships 163–170
Spearman's rank correlation
    coefficient 114–116
standard deviation
  Excel 31–33
  exercises 33–35
  probability 39–45
  summarizing data
    numerically 30
  worked example 31–33

standard error of the intercept,
    linear regression   137
standard error of the mean
    46, 47
    exercises   47
standard error of the slope, linear
    regression   137
standard form *see* scientific
    notation
standardized value *see* z-score
stationarity
    assumptions   180
    time series analysis   179–180
statistical tables   205–239
    critical values of the
        F-distribution   212–233
    critical values of the
        U-distribution   234–237
    critical values of the $\chi2$
        distribution   238–239
    t-distribution   210–211
    z-table   206–209
stratified sampling   17
    bias   17
summarizing data numerically
    24–33
    central tendency of datasets
        24–28
    coefficient of variation
        30–33
    dispersion   28–29
    kurtosis   33
    skewness   33
    standard deviation   30
    variance   29–30
systematic sampling   16

t
target population   14
t-distribution
    cf. normal distribution   51–52
    confidence intervals   50–53

t-distribution (statistical table)
    210–211
time series analysis   173–192
    assumptions   180
    autocorrelation   189–190
    correlograms   190, 191
    decomposing time
        series   179–180
    definitions   174–175
    exercises   190, 192
    geographical
        research   173–174
    irregular
        fluctuations   179–180
    moving average   181–186,
        187
    R statistical
        software   178–179
    seasonality   179–180
    seasonal variation   187–189
    stationarity   179–180
    time series plots   21–22,
        175–179
    trends   179–186
transverse Mercator
    projection   151–152,
    154, 155
trends, time series
    analysis   179–186
t-test   61–70
    comparing means   61–70
    Excel   67
    paired samples (paired
        t-test)   71–73
    R statistical software   67–68
two-way analysis of
    variance   107
Type I error (false positive)
    correlation and causality   117
    hypothesis testing   57–58
Type II error (false negative),
    hypothesis testing   57–58

u

uncertainties, reporting data
and   7–9
Universal Polar Stereographic
(UPS) projection   151, 153
UTM (Universal Transverse
Mercator) system   151,
152–153

v

variables, hidden, correlation and
causality   116–117
variance
analysis of variance (ANOVA)
89–108
degrees of freedom   30
Excel   31–33
summarizing data numerically
29–30
worked example   31–33
vector data   5
vectors, R statistical software
196–197

w

weighted mean centre, spatial
data   157–158

Wilcoxon rank sum test
78–79
see also Mann–Whitney *U*-test
worked examples
central tendency of datasets
26–28
coefficient of variation   31–33
frequency distributions
19–21
scientific notation (standard
form)   12
standard deviation   31–33
variance   31–33
World Geodetic System 1984
(WGS84) reference ellipsoid,
map projections   150
writing data, R statistical
software   202–204

z

z-score   47
normal distribution   39–42
probability   39–42
z-table (statistical table)
206–209
Z-test, hypothesis testing
56–61

Printed and bound by CPI Group (UK) Ltd, Croydon, CR0 4YY

27/10/2024

14580368-0001